SCIENCE
and
RELIGION

SCIENCE
and
RELIGION

Are They Compatible?

edited by

PAUL KURTZ

with the assistance of
BARRY KARR *and* RANJIT SANDHU

Published for
The Center for Inquiry

 **Prometheus Books**

59 John Glenn Drive
Amherst, New York 14228-2119

Inquiries should be addressed to
Prometheus Books
59 John Glenn Drive
Amherst, New York 14228–2119
VOICE: 716–691–0133, ext. 210; FAX: 716–691–0137
WWW.PROMETHEUSBOOKS.COM

11 10 09   12 11 10 9

Library of Congress Cataloging-in-Publication Data

Science and religion : are they compatible? / edited by Paul Kurtz ; with the
   assistance of Barry Karr and Ranjit Sandhu.
      p. cm.
   Includes bibliographical references.
   ISBN 978–1–59102–064–6 (pbk. : alk. paper)
      1. Religion and science. I. Kurtz, Paul, 1925– II. Karr, Barry. III. Sandhu,
Ranjit.

BL240.3.S346 2003
291.1'75—dc21

                                                                2003041378

Every attempt has been made to trace accurate ownership of copyrighted material in this book. Errors and omissions will be corrected in subsequent editions, provided that notification is sent to the publisher.

Printed in the United States of America on acid-free paper

CONTENTS

Introductory

I. Cosmology and God

124535

II. Intelligent Design: Creationism versus Science

III. Religion and Science in Conflict

IV. Science and Ethics: Two Magisteria

V. The Scientific Investigation
of Paranatural Claims

VI. Scientific Explanations of Religious Belief

VII. Accommodating Science and Religion

Introductory

1

AN OVERVIEW OF THE ISSUES

PAUL KURTZ

I

Is science compatible with religion, or must they of necessity conflict? Their relationship has long been debated; and pitched battles of varying degrees of
intensity have been waged throughout history between the partisans of these
two areas of human interest. Although science had its precursors in ancient
Greece and Rome, the dawn of modern science began in the sixteenth and seventeenth centuries, when scientists and philosophers developed a new method
of inquiry. These pioneers abandoned tradition, mysticism, revelation, and
faith, and proceeded directly to the Book of Nature. They eschewed hidden
occult explanations for natural causes. They rejected purely speculative metaphysics and sought hypotheses and theories that were verified by empirical
observation, experimental prediction, and the precision and power of mathematics. Scientific principles were considered tentative, open to revision in the
light of new data or more comprehensive theories.

Scientific progress could only occur when the theological and philosophical authorities of the past were discarded, and a fresh bold approach to nature
was adopted. This led to unparalleled breakthroughs in field after field of
research. Beginning with physics, astronomy, and the natural sciences, scientific

methods of inquiry were later extended in the nineteenth century to chemistry, biology, and the life sciences, and eventually in the nineteenth and twentieth centuries to the social and behavioral sciences. As scientific research advanced the frontiers of knowledge, the findings of basic research were applied to technology with extraordinary results, impacting civilization enormously and transforming moral values and political, economic, and social institutions.

Many theologians reacted vociferously, fearful that science was encroaching on their domain; they fought back, but this proved to be only rearguard defensive skirmishes, for scientific research continued, impressing everyone by its discoveries. Still, scientists were constantly confronted by threats of censorship by religious and political authorities. Ever cognizant of the long history of inquisitions and heresy trials, they sought to defend the integrity and freedom of research, which their critics found difficult to prohibit in the face of its evident achievements.

Unfortunately, these conflicts continue today in various parts of the globe, though large sectors of the religious establishment, at least in the West, now accept the principles of free inquiry and are receptive to the findings of science (no doubt for their economic value), and indeed encourage its growth and development. Nonetheless, there are still shrill religious opponents—such as the fundamentalists who oppose Darwinian evolution in the United States and militant Muslims in the Middle East and throughout the world who oppose secular democracy. Some orthodox religionists still harbor misgivings about scientific inquiry and its alleged deleterious influence on traditional beliefs and values.

Religious doctrines and institutions predate the growth of modern science. Christianity, Judaism, Islam, Hinduism, Buddhism, and other religious traditions have deep roots in human history—sacred books; revered teachers and saints; the majesty of the arts; resplendent cathedrals, temples, and mosques; and deep philosophical traditions are imbued with a cultural heritage that emphasizes the need for faith and devotion. These religious institutions are ingrained in the very fabric of human civilization—its languages, concepts, and values; they define who and what we are. Indeed, they are among the oldest institutions in human history, enduring constant challenges to their hegemony. Religious prophets have been deified for what they promise. They offer meaning and purpose to life and promises of eternal salvation from this vale of tears. These are based on supernatural sources outside of nature. Science, it is claimed, deals with nature, but the divine reality transcends this in every way over and beyond space and time. Religions have had a deep effect on all social institutions, education, politics, the economy, and the law. The great religions that have survived have grown by osmosis. They are embraced by individuals who lived (until recently, at least) in demarcated geographical areas. A person's religious convictions were usually a condition of birth and not of choice; though during periods of change, where sects contend or invading armies bring a new faith, there may be conversions and deconversions.

Inculcating the young into its rich religious heritage is essential for its perpetuity. Rules of marriage and reproduction and rituals governing all aspects of life reinforce the power and influence of the Creed. Historically, to question sacred doctrine was to shake the foundations of the social order. Thus religions have persisted because they were indoctrinated by custom and habit, sustained by law and rooted in faith and feeling; unexamined, they were simply assumed to be true, and defended as encompassing Absolute Truth and Virtue. Blasphemy and iconoclasm, even religious dissent, during the long history of humankind were considered to be both sinful and subversive and were punishable by excommunication, ostracism, or death. In theocratic societies, the power of secular rulers is invoked to enforce religious dominance.

There is a profound difference between science and religion in its conception of truth. Science requires an open mind, free inquiry, critical thinking, the willingness to question assumptions, and peer review. The test of a theory or hypothesis is independent (at least one would hope) of bias, prejudice, faith, or tradition; and it is justified by the evidence, logical consistency, and mathematical coherence. Science claims to be universal (though postmodernist critics deny this), transcending specific cultures and replicable in any and every laboratory in the world. Although religions claim to be universal, they have split into contending factions concerning hegemony: they rely on the acceptance of faith in specific revelations and their interpretation by differing prophets, priests, ministers, rabbis, monks, or mullahs.

Religions have been challenged in the Western world by philosophers and scientists from the earliest days in Greece and Rome: Socrates, Plato, Aristotle, the Epicureans, and Skeptics lambasted the reigning mythologies of their day, recommending Reason as the guide. Christianity in the Western world eventually overwhelmed the schools of philosophy and the pagan religions and supplanted them, especially after the Emperor Constantine declared

Christianity to be the official religion of the Roman Empire. Augustine, in the fifth century C.E., rejected philosophical skepticism and defended faith in God as our primary duty, and the City of God came to dominate the City of Man for almost a thousand years.

A radical departure occurred with the European Renaissance beginning in the fourteenth and fifteenth centuries: This was made possible because of the rediscovery of the classics, especially the works of Aristotle, which had been lost to the West and preserved by Islamic scholars. Indeed, there was a brilliant period in the twelfth century in Cordoba, Spain, when Averroës (1126–1198), drawing upon the texts of Aristotle, argued for the autonomy of philosophy and made possible their translation into Arabic, Latin, and Hebrew. The writings of Averroës were burned by a fanatic mob a year before he died. After his death, Islamic scholars rejected his view about the autonomy of philosophy, and conservative Muslim theologians, quoting Al-Ghazzali (1058–1111), suppressed further efforts at free inquiry and defended the preeminence of the Koran and Hadith.

The works of Aristotle were transmitted to western Europe. Thomas Aquinas (1224–1274) worked out a new philosophical and theological system which left room for both reason (natural theology) and revelation (revealed theology). Aquinas thought faith in revelation was consonant with the principles of reason. Thus the scientific and metaphysical system of Aristotle became dominant in the High Middle Ages and was considered to be in accord with Judaic-Christian scriptures.

In the modern world, the writings of the skeptics were rediscovered, casting further doubt on the authority of the church over science. Natural philosophers and physicists appealed to mathematics, observation, and experiment, not faith or revelation, to demonstrate the newly formulated laws of mechanics. This led to the overthrow of the Thomistic-Aristotelian system of thought. The first major conflict between science and religion, at least its most dramatic forms, emerged when Giordano Bruno (1548–1600) was burned at the stake and Galileo Galilei (1564–1642) was placed under house arrest for doubting received doctrine.

The question was raised, What happens when scientific discovery contradicts the authority of the church? Which should prevail? This challenge reached a critical boiling point in astronomy when the heliocentric Copernican theory, reinforced by Galileo, Kepler, and Newton, placed the Sun rather than the Earth at the center of the solar system, and rejected the Ptolemaic system of epicycles. A deterministic, mechanical, and material world replaced the biological framework of potentiality, actuality, and teleology of Aristotle.

The conflict between religion and science continued on other fronts. The Protestant Reformation, appealed to freedom of conscience of the individual rather than church authority, which was often allied with ancien régimes. Democratic revolutions demanded toleration for dissenting viewpoints. The Enlightenment of the eighteenth century extended the methods of science—understood as the combination of rationalism, empiricism, and experimentalism. They sought to apply science to all areas of human interest, including the political economy, social studies, and psychology—all for the betterment of humanity. *Les Philosophes*, such as Condorcet, thought that science could refashion ethics and reconstruct society, thus liberating human beings from the dead hand of tradition and contributing to the progressive improvement of humankind. The emancipation of humanity from entrenched hierarchies became a rallying cry for the American and French revolutions, which heralded liberty, equality, and fraternity.

In the nineteenth century Darwinism precipitated an intense battle between religion and science; for it overthrew the classical view that there were fixed species; and it postulated natural causes for the evolution of life, including the descent of the human species from other primates. From a purely scientific standpoint, there was no need for the creationist theory and it was replaced by evolutionary hypotheses. This heated controversy continues today. Many conservative religionists seek either to supplant evolutionary theories, or at least to have them supplemented by theories of intelligent design. In the United States they have sought to limit or modify the teaching of evolution in the schools. The point of contention is whether science requires secularism—the separation of church and state—or whether theological doctrines can be appealed to in order to censor scientific inquiry. The effort by established churches to define orthodoxy and punish blasphemy has largely been defeated in Western democracies, though it is still a burning issue in Islamic lands.

Warfare between religion and science continues today in many fields of inquiry, especially where basic moral values are held to be at stake. For example, the issue of reproductive freedom for women has engendered intense political controversy. Contraception and abortion are still at the center of the debate, in medical science, as are the right to die with dignity, euthanasia, and assisted suicide. Shall the theological doctrine of an independent, "immortal soul," created by God, serve as a basis for opposition to abortion, birth con-trol, and artificial insemination, and/or the right to control one's own body? This view spills over into other areas of biomedical research: should cloning, even therapeutic cloning that uses stem cells, be permitted? No, say conserva-tive theologians; yes, say many research scientists. Similar issues have been raised on moral grounds concerning the genome project and biogenetic engi-neering. Some theologians maintain that biogenetic research should not be permitted to transform human nature—which has been endowed by God. Proponents respond that we intervene all the time in the evolutionary process, for society protects the weak and handicapped who might normally die, and that in any case, the utilitarian benefits of research to the health and welfare of humans far outweighs the objections.

This raises the question of whether theology can or should censor scien-tific inquiry in the name of a higher morality. During the Cold War many con-sidered science evil for opening the Pandora's box of nuclear-energy research; and there was widespread apprehension that an Armageddon doomsday sce-nario could destroy all life on the planet. This raised the basic question of the relationship of ethics to science. Should moral questions be left primarily under the domain of religion? Or can science assist humankind in making wiser moral judgments, by understanding the causes of human behavior, by inventing new means to achieve our ends, and by testing normative recom-mendations in part by their observed consequences?

Serious moral differences emerge in society as to which forms of research should be funded and pursued. Questions of value intervene in science con-tinuously: for example, many applied scientific analysts today challenge poli-cies on population growth, global warming, the depletion of natural resources, the despoliation of the environment, or even the role of women or children in patriarchal societies. Should the free market and/or the democratic ballot decide these issues, or should decision making on the national or global level take scientific recommendations into account? If so, what values should be controlling? Can a theological framework rule certain issues out-of-bounds? The interface between scientific research and ethics necessarily involve politi-cal-economic considerations, but these often rest on theological-moral assumptions. The advance of naturalistic science not only threatens traditional supernatural perspectives based on ancient literatures, but it is likely to funda-mentally alter ethical values as cultures collide and migration, invasion, and war bring new multicultural lifestyles and opposing value systems. Technological innovation constantly unsettles traditional morality. Science is viewed in many parts of the world as a revolutionary force threatening the hegemony of religion. Backward nations desperately wish to use the fruits of technology to raise the standard of living of their populations; at the same time many fear the erosion of traditional religious values.

Thus the conflict between modern science and religion today assumes new forms. Some Western postmodernists and multiculturalists seek an accommodation by insisting that science is only "one mythology among others": they affirm that science is relative to culture and that objective truth is an illusion and need not supplant traditional religions. In response the supporters of science maintain that it is in touch with the real world, for its hypotheses are tested experimentally. Thus it provides us with universal or general knowledge that is global in reach and transcends the limited perspective of any particular culture. The conflict between science and religion thus takes on new and compelling forms today, for it is related to what Samuel Huntington has characterized as "a clash of civilization." Is there any way for science to resolve these conflicts or are religion and ethics entirely beyond its domain?

No doubt the deeper question concerns the nature of truth. Are there two truths or two cultures: scientific inquiry *versus* religious faith? Or is there only one method of inquiry and one kind of truth? Naturalists today are committed to using the methods of scientific inquiry: they maintain that the quest for natural causes is the most appropriate purpose of inquiry, and they wish to extend science to all areas of human interest. Many religionists proclaim that there are higher forms of religious truths, not amenable to science, and that these are discovered by faith and revelation.

II

On the current cultural scene, there is a vocal chorus singing praises to the alleged mutual harmony and support of science and religion. The claim is made that there are two realms or magisteria, each with a vital role to play. Many skeptics have serious misgivings about this alleged rapprochement, though some are reluctant to criticize religion openly. The question is thus raised: to what extent can or should skepticism be applied to religious claims?

By the term *skepticism* I do not refer to the classical philosophical position which denies that reliable knowledge is possible. Rather, I use the term to refer to *skeptical inquiry*. There is a contrast between two forms of skepticism: (1) that which emphasizes doubt and the impossibility of knowledge, and (2) that which focuses on *inquiry* and the genuine possibility of knowledge; for this latter form of skepticism—"the new skepticism" as I have labeled it—skeptical inquiry is essential in all fields of scientific research.

What I have in mind is the fact that scientific inquirers formulate hypotheses to account for data and solve problems. Their findings are tentative. They are accepted because they draw upon a range of confirming evidence and predictions and/or fit into a logically coherent theoretical framework. Reliable hypotheses are adopted because they are corroborated by a community of inquirers and because the tests that confirm them can be replicated. Scientific hypotheses and theories are fallible, and in principle they are open to question in the light of future discoveries and/or the introduction of more comprehensive theories. The point is that we have been able to achieve reliable knowledge in discipline after discipline because of the effective application of skeptical inquiry. Now the central questions that have been raised concern the range of skeptical inquiry. Are there areas such as religion in which science cannot enter?

Science has always had its critics, who have insisted that one or another area of human interest is immune to scientific inquiry. At one time it was proclaimed that astronomers could never know the outermost reaches of the universe (August Comte), the innermost nature of the atom (John Locke), or human consciousness (Henri Bergson). Critics have also insisted that we could not apply science to one or another aspect of human behavior—political, economic, social, or ethical behavior; the arts; human psychology; sexuality; or feeling. I do not think that we should set a priori limits antecedent to inquiry; nor should we seek to denigrate the ability of scientific investigators to extend the frontiers of research into new areas.

Specifically, should skeptical scientific inquirers question the regnant sacred cows of religion? There are both theoretical and prudential issues here at stake. I can find no theoretical reason why not, but there may be practical considerations. For one, it requires an extraordinary amount of courage today as in the past (especially in America!) to criticize religion. One can challenge politicians, celebrities, cult figures, paranormal hucksters, mediums, psychics, and astrologers with abandon, but to question the revered figures of orthodox religion is another matter, for this may still raise the serious public charge of blasphemy and heresy; and this can be dangerous to one's person and career.

History vividly illustrates the hesitancy of skeptics to apply their skepticism to religious questions. In ancient Rome, Sextus Empiricus, author of *Outlines of Pyrrhonism*, defended the suspension of belief in regard to metaphysical, philosophical, and ethical issues. He did not think that reliable knowledge about reality or ethical judgments was possible. He neither affirmed nor denied the existence of the gods, however, but adopted a neutral stance. Since there was no reliable knowledge, Pyrrho urged that compliance with the customs and religion of his day was the most prudent course to follow. The great skeptic Hume bade his friend Adam Smith to publish his iconoclastic *Dialogues Concerning Natural Religion* after his death (in 1776), but Smith declined to do so, disappointing Hume. Hume's nephew David arranged for posthumous publication. The French author Pierre Bayle (1647–1706) perhaps expressed the most thoroughgoing skepticism of his time. In his *Dictionnaire historique et critique*, Bayle presented a scathing indictment of the prevailing theories of his day, finding them full of contradictions. He was highly critical of religious absurdities. He maintained that atheists could be more moral than Christians and that religion did not necessarily provide a basis for ethical conduct. Nonetheless, Bayle professed that he was a Christian and a Calvinist, and this was based upon pure faith, without any evidence to support it—this is known as fideism. Did Bayle genuinely hold these views or was his fideism a ruse to protect his reputation and his fortune?

This form of fideism, I maintain, on theoretical grounds is illegitimate, even irrational. For if, as skeptical inquirers, we are justified in accepting only those beliefs that are based upon evidence and reason, and if there is no evidence either way or insufficient evidence, should we not suspend judgment, or are we justified in taking a leap of faith? If the latter posture of faith is chosen, one can ask, On what basis? If a person is entitled to choose to believe whatever he or she wishes, *solely* or *largely* because of personal feeling and taste, then "anything goes." But this anarchic epistemological principle can be used to distort honest inquiry. (The implication of this argument is that if you do not have a similar feeling, you are entitled *not to believe*.) One may ask, Can one generalize the epistemological rule; and if so, is someone entitled to believe in angels, gremlins, vampires, or demons on the basis of feeling and fancy? The retort of the skeptical inquirers is that where there is insufficient evidence to decide the question, we are not justified in believing, though there may be prudential reasons in society where individuals decide to remain silent.

But as a matter of fact, most of those who believe in the traditional religions do not base it on pure fideism alone, but on reasons and evidence. Indeed, no less an authority than Pope John Paul II maintained the same in an encyclical entitled *Faith and Reason*. In this, the pope condemns both fideism and atheism. He attacks, for example, the naive faith in "astrology and the New Age." He criticizes "exaggerated rationalism" and pragmatism on the one hand and postmodernism on the other, but he also condemns the exclusive reliance on faith. The pope maintains that reason and scientific inquiry support rather than hinder faith in Christian revelation and Catholic doctrine. Skeptics might agree with the pope's defense of reason and scientific inquiry, yet question whether these do indeed support his own beliefs.

Thus for the skeptical inquirer acquiescence to the fideist's rationalization for his beliefs is mistaken. Similarly, in answer to those theists who maintain that there is adequate evidence and reasons for their belief, skeptical inquirers should not simply ignore their claims, saying that they are beyond scientific confirmation, but should examine them. Since the burden of proof is always upon the claimant, skeptical inquirers may question both the fideist and the partial-evidentialist in religion, if they do not believe that they have provided an adequately justified case.

The upshot of this controversy is that scientific and skeptical inquirers, in theory at least, should deal with religious claims. Not to do so is to flee from an important area of human behavior and interest and is irresponsible. By the same token, skeptical inquiry in principle should apply equally to economics, politics, sociology, ethics, and indeed to all fields of human interest. From this viewpoint science should not be so narrowly construed that it only applies to the natural and biological sciences; it should bring in the tools of experimental inquiry, logical analysis, historical research, and rational investigation to the social and behavioral sciences. In this sense religious claims would be amenable to scientific examination and skeptical inquiry.

III

With this in mind, the Center for Inquiry (supported in its endeavor by both the Committee for the Scientific Investigation of Claims of the Paranormal, publisher of the *Skeptical Inquirer*, and the Council for Secular Humanism, publisher of *Free Inquiry*) decided to convene a special conference in Atlanta, Georgia, in November 2001, on the theme "Science and Religion: Are They

Compatible?" The Center for Inquiry is committed to science, reason, and freedom of inquiry in all areas of human interest. It does not wish a priori to limit scientific inquiry or denigrate its possible achievements.

Many conferences and symposia have been convened in recent years, devoted to exploring the relationship between science and religion. The Templeton Foundation, which is sympathetic to a religious-spiritual perspective, in particular has funded many such conferences. Most of these have begun with the premise that religion and science are mutually compatible, reinforcing each other. This has been inspired, in part, in the United States because of the growth of religiosity, the power of religious lobbies, and the desire of religionists to find a place for religion within the scientific worldview. This has been brought to the fore because everyone has been impressed by the continuous advances of scientific inquiry on the frontiers of knowledge, and the evident impact of technology on society, the state, and the economy. Does this threaten to undermine religion entirely, or can a modus vivendi or an accommodation be worked out?

We invited scientists and scholars predominantly skeptical of the paranormal and religion to this three-day conference, though other speakers sympathetic to theistic religions were also invited to engage in debate (some declined our invitation). Basically, skeptical inquirers until now have been all but drowned out by the din of proreligious scientific believers, so a skeptically oriented scientific conference seemed appropriate to convene.

This book includes several of the talks delivered at that conference, but it also draws on articles published in the *Skeptical Inquirer*, which devoted two special issues to Science and Religion, and *Free Inquiry*, which is committed to the naturalistic and secular viewpoint as it applies to religion and society, and seeks rational ethical alternatives based on the sciences.

The contributors to this volume take a new approach. Most are committed to methodological naturalism; namely, the principle that claims to knowledge should be submitted to the tests of scientific inquiry, and that we should not exclude anything a priori from its purview. The papers in this volume cut across disciplines. These include articles by physicists, astronomers, mathematicians, biologists, botanists, zoologists, anthropologists, psychologists, cognitive-studies inquirers, historians of science, and philosophers.

Part I, "Cosmology and God," discusses the implications of the new scientific cosmologies for religion. Do they leave room for God? Steven

Weinberg, well-known Nobel Prize astronomer, denies that they do and he presents the scientific case against Design. Physicist Victor J. Stenger, author of many books critical of religion and the paranormal, deals with "The Anthropic Principle." Many of its proponents maintain that the universe is so "fine-tuned" to the emergence of life, particularly human life, that this suggests a kind of design or purpose. David Shotwell, a mathematician, says that the anthropic principle really invokes supernaturalism in science, which is unwarranted. Harvard astronomer Owen Gingerich believes that the current cosmological outlook in physics is consonant with Christianity, which he accepts. Quentin Smith, a philosopher of science, disputes the view that the Big Bang cosmological outlook provides evidence for theism. He argues that an atheistic interpretation is perfectly amenable with quantum mechanics and that the universe may very well be due to a random quantum fluctuation. Astrophysicist Neil deGrasse Tyson critically discusses the "Holy Wars" now underway and the God question from the standpoint of the scientific naturalist.

Part II, "Intelligent Design: Creationism *vs.* Science," brings to the fore one of the most controversial issues on the current scene: Is the Darwinian-inspired evolutionary theory mistaken, and is intelligent-design a more appropriate model? *Skeptical Inquirer*'s distinguished editor Kendrick Frazier focuses on the political implications for scientific inquiry of creationism and intelligent design in America in its battle against evolution. Philosopher-theologian William A. Dembski is a defender of the traditional design argument and thinks that skepticism has few prospects for unseating this approach. Massimo Pigliucci, an evolutionary geneticist, disagrees; intelligent design has no experimental significance for current evolutionary theory, he says, and is only a mask for traditional creationism. Anthropologist Eugenie C. Scott, considered to be the leading critic of creationism in its battle against science, evaluates the new "science-and-religion movement." Will it succeed in finding an accommodation between science and religion? Last, Taner Edis, a young physicist who was born in Turkey and emigrated to the United States, shows how creationism has become a serious force in undermining science in the Islamic world.

Part III, "Religion and Science in Conflict," generally is skeptical that an accommodation can be easily reached between religion and science. Historian Vern Bullough places the battle between religion and science in its broader historical context. Timothy Moy reexamines the Galileo affair. This is a symbolic case in the defense of freedom of scientific inquiry against religious repression—though he believes that the Catholic Church's position has been misinterpreted. Sir Hermann Bondi (of the Bondi-Hoyle-Gold hypothesis of "constant creation") maintains that science strives for universality, whereas religion is divisive. The article by Daniel Dennett lays forth the epistemological case for "getting it right" by using science to test truth claims, and he rejects multicultural postmodernism. Jacob Pandian argues that the quest for cooperation between religion and science can be dangerous insofar as it undermines the scientific endeavor. Similarly, Barry Palevitz defends science in its dispute with religion. He expresses concern that his students press him to confess a religious belief. The section is capped off with a piece by Arthur C. Clarke, the famous science-fiction author of *2001*, presenting his scientific "Credo" in contrast with religious superstition.

Part IV, "Science and Ethics: Two Magisteria," enters into the fray by asking, What is the relationship between ethics, science, and religion? Stephen Jay Gould, the noted zoologist and defender of Darwinian evolution, argues in a highly controversial piece that there are "two nonoverlapping magisteria," that of science, which is concerned primarily with knowledge about the world and nature, and that of religion, which he thinks has an appropriate role in ethics. Richard Dawkins, the equally forceful public spokesperson for science, disputes Gould and argues that there are irreconcilable differences between science and religion which cannot be glossed over. He criticizes the effort by theologians to interpose "the soul" in the evolutionary process. Richard Feynman, Nobel Prize physicist, points out that the traditional metaphysical views of religion are undermined by contemporary science. He agrees, almost as an afterthought, with Gould that ethics is an appropriate domain of religion.

Part V, "The Scientific Investigation of Paranatural Claims," begins with an article in which I propose that supernatural effects, insofar as they are immanent in the world, can be examined scientifically. I introduce the term *paranatural*, illustrating how empirical tests can be applied by science to the religious claim that there is life after death. Psychologists Richard Wiseman and Ciarán O'Keeffe apply rigorous scientific protocol to appraise after-death-communications studies. Antony Flew, well-known British analytic philosopher, examines "near-death experiences." He is doubtful of the concept of a "separable soul" and raises the question in at least one case of alleged empirical confirmation of whether the soul has left the body. Physicist Jerome W. Elbert discusses the conflict between the historic religious conceptions of the soul and the contemporary scientific approach, which dispenses with the concept entirely. Irwin Tessman and Jack Tessman evaluate the recent claim that there is empirical evidence for the "efficacy of prayer." Joe Nickell shows how the scientific investigation of the Shroud of Turin indicates that it is most likely a fraud from the thirteenth century. He interprets "Shroud Science," which affirms its authenticity, as a form of pseudoscience.

Section VI, "Scientific Explanations of Religious Belief," begins with my article in which I focus on the question, Why do people believe or disbelieve? I contrast the role of human cognition with possible genetic biological determinants of religious belief. Anthony Layng maintains that religions are a product of cultural evolution. Morton Hunt examines the sociobiological roots of religion, maintaining that religious beliefs and practices were adaptive during the evolutionary history of the human species. Steven Pinker further examines the question whether religious beliefs are due to cultural conditioning or genetic dispositions. David C. Noelle is skeptical of the claim that human beings are wired to believe in God.

Section VII, "Accommodating Science and Religion," begins with a provocative essay by Martin Gardner on "Science and the Unknowable." Gardner, one of the leading skeptical critics of paranormal and religious cults, nonetheless remains a convinced fideist. In this essay he maintains that the range of the unknowable is so vast, including possible multiverses, that this inspires in him awe, wonder, mystery, and a sense of piety. James Lovelock, the author of *Gaia*, the thesis that the Earth is like a living organism and should be revered by humans, believes that devotion to environmental preservation of the planet provides a moral credo for agnostics. Chet Raymo brings to the fore a kind of spiritual interpretation of nature, though it is naturalistic in character. Matt Young believes that science and religion are incompatible, and that a humanistic and naturalistic approach is perfectly reasonable.

In "Afterthoughts," I argue that religion is incompatible with science if it seeks to compete with it. Religion cannot substitute its truths for those of science. Religion should neither seek to impose its theological doctrines in the political arena nor dominate ethics. I conclude, however, that science and religion *are* compatible—but only if religion is reinterpreted primarily as a form of existential poetry, dramatizing the fragility and contingency of the human condition in an impersonal universe and recognizing with awe and wonder the vastness and mystery of the cosmic scene.

2
ARE SCIENCE AND RELIGION CONFLICTING OR COMPLEMENTARY?
Some Thoughts About Boundaries

KENDRICK FRAZIER

Science and religion are two great manifestations of human culture. To what degree are they complementary and to what degree are they in opposition? Does recognition of their separate dominions require mutual distrust or is there room for consilience that does injustice to neither? How can the great issues that separate them best be handled? What about the slightly differing approaches to these issues by what might be termed science-minded skeptics (to simplify, rationalists motivated mainly by an appreciation of science and reason whatever their views about religion) and humanists (rationalists motivated by those same qualities plus a resistance to the idea of a god)? Do they likewise have grounds for mutual agreement that can take strength from their broad philosophical commonalities without splintering over their one essential difference? What role should pragmatic questions (such as recognition of the immense religiosity of American culture) play in these considerations if you are to maintain intellectual integrity? If you are to be effective? While there is really only one kind of science, there are multitudes of religions that span broad spectrums of belief and

Kendrick Frazier's "Are Science and Religion Conflicting or Complementary? Some Thoughts About Boundaries" originally appeared in the *Skeptical Inquirer* 23, no. 4 (July/August 1999).

of degrees of rigidity versus openness toward science and reason. Can mainstream religions that welcome science be helpful in mitigating the attacks of fundamentalist religions that oppose it?

For scientists and science-minded skeptics examining claims to knowledge, the most troublesome boundary issue of all is that between science and religion. Likewise skeptics and humanists have a bit of a boundary problem themselves; they have enormous areas in common but experience some unease over this one interface where they differ. Some data: Polls show that 90 percent of the American people describe themselves as religious. This is much higher than in many of the other countries of the Western world. Yet, perhaps surprisingly, of Americans in the twelve largest Christian denominations, 89.6 percent belong to churches that support evolution education—a subject where science and religion are often perceived to be in disagreement. It is easy to stereotype religious belief, but the situation is far more complex and multilayered than is usually portrayed.

As a journal of science and reason, the *Skeptical Inquirer* has generally limited itself to those questions in which some kind of testable, empirical claim has been made. (In the areas of religion, this includes the claims of "scientific" creationism and its attacks on evolution, factual assertions about the Shroud of Turin, testable claims of modern "miracles" such as weeping icons, "Bible code" predictions, and the like.) This has served us well. It is appropriate because science itself is concerned with empirical questions; science formulates explanations of the natural world on the basis of natural, rather than supernatural, causes. All science operates with the view that supernatural causes are both unnecessary and unfruitful in formulating hypotheses about the natural world. The *Skeptical Inquirer* reflects and expresses that scientific viewpoint.

Assertions about ultimate causes and questions, such as whether there's a God or an afterlife, are of interest to much of humanity and appropriate for philosophical and theological consideration. But they are generally considered to be outside the scope of science. It isn't that such questions aren't important; it is that science, most properly and typically defined, has no ability to resolve them. Science most of all is pragmatic; unfalsifiable questions offer no hope of yielding scientific answers.

Many scientists and skeptical inquirers dispute this more limited interpretation and wish to push science more strongly into criticism of the realms of belief. They hold not only that the laws of nature are sufficient to explain the natural world but that the supernatural does not exist and we should be forthright about that. This view has a long and respected tradition within philosophy, but it is a philosophical one rather than a scientific one, and that distinction should be recognized.

In addition to strictly scientific questions, however, the *Skeptical Inquirer* has always considered the wider philosophical, psychological, social, and educational implications of beliefs of all kinds. This is the case whether they are based in logic and reason, the "paranormal," mysticism, pseudoscience, fringe-science, or religion. Our essential interest is in the advancement and understanding of science, in the use of reason and rationality, in free intellectual inquiry and dispassionate examination of claims to knowledge, and in deeply understanding all important social beliefs and trends. We have no interest in ideology or in proselytizing—only in promoting the use of the diverse methods of science and in applying high standards of scientific inquiry, evidence, reason, and logic.

In this sense some questions and issues along the borders of science and religion are appropriate for our consideration. But it is a fine line that needs to be walked carefully, and we have always tried to do so. With 90 percent of the population in the United States self-described as religious (compared with about 50 percent who in surveys express belief in some aspects of the paranormal), this broadly religious social milieu must be recognized as a reality. If arguments based on science and reason are to be heard sympathetically by the wider public you wish to influence, they should be presented with sensitivity toward those who may hold beliefs other than your own. This is merely a matter of respect, sticking to facts rather than opinions, and avoiding any appearance of arrogance or superiority. It is counterproductive to act in any other way. Second, while science deals with facts, evidence, and hypotheses rather than belief, a person's right to hold his or her own beliefs—contrary to your own or even contrary to scientific evidence—must be acknowledged. At the same time any efforts to force those personal beliefs on others, bring them into the science classroom in any guise, or pretend that one's religious beliefs are supported by science must be firmly resisted.

Science is concerned with understanding the natural world, religion with humanity's moral, ethical, and spiritual needs. Were it such a neat division! If science and religion kept to these separate domains, there would be no conflict. However, overlaps or intrusions of one realm into the other are common and inevitable. New scientific discoveries and technological advances both affect our lives in many ways, both good and bad, and continually raise new issues of ethics and human values. Some commingling of science and religion is even encouraged by well-meaning people who find deep value in both and understandably seek to bring them into some kind of unification. They desire to have no compartmentalized aspects of personal philosophy.[1] They rightly point out that science and religion both address deep questions about the world and our place in it. Both arise in part from curiosity and awe about the world. So to some degree consilience may be possible, but only by clearly recognizing the great differences between science and religion.

Science (and reason) must not yield any of its own ground. Science is based foremost on evidence, not authority or revelation. In science, nothing is taken on faith, while in religion, faith is at the heart of belief. In science all knowledge is tentative, continually subject to revision when better explanations and evidence (always aggressively sought) are acquired; religion asserts the presence of unchanging and unchallengeable eternal truths. Science proposes explanations about the natural world and then puts those hypotheses to repeated tests using experiments, observations, and a creative and diverse array of other methods and strategies. Many religions discourage skepticism or critical examination of cherished precepts. This commitment to test the validity of ideas and claims separates science from religion.

Note

1. This desire to have no compartmentalization is also true, of course, of humanists, who understandably see their opposition to religion and their criticisms of the paranormal as a singular part of the same personal philosophy.

I

Cosmology and God

3
A DESIGNER UNIVERSE?

STEVEN WEINBERG

I have been asked to comment on whether the universe shows signs of having been designed. I don't see how it's possible to talk about this without having at least some vague idea of what a designer would be like. Any possible universe could be explained as the work of some sort of designer. Even a universe that is completely chaotic, without any laws or regularities at all, could be supposed to have been designed by an idiot.

The question that seems to me to be worth answering, and perhaps not impossible to answer, is whether the universe shows signs of having been designed by a deity more or less like those of traditional monotheistic religions—not necessarily a figure from the ceiling of the Sistine Chapel, but at least some sort of personality, some intelligence, who created the universe and has some special concern with life, in particular with human life. I expect that this is not the idea of a designer held by many here. You may tell me that you

Steven Weinberg's "A Designer Universe" is based on a talk given in April 1999 at the Conference on Cosmic Design of the American Association for the Advancement of Science in Washington, D.C. It was first published in the *New York Review of Books* (October 21, 1999) and later in the *Skeptical Inquirer* 25, no. 5 (September/October 2001).

are thinking of something much more abstract, some cosmic spirit of order and harmony, as Einstein did. You are certainly free to think that way, but then I don't know why you use words like "designer" or "God," except perhaps as a form of protective coloration.

It used to be obvious that the world was designed by some sort of intelligence. What else could account for fire and rain and lightning and earthquakes? Above all, the wonderful abilities of living things seemed to point to a creator who had a special interest in life. Today we understand most of these things in terms of physical forces acting under impersonal laws. We don't yet know the most fundamental laws, and we can't work out all the consequences of the laws we do know. The human mind remains extraordinarily difficult to understand, but so is the weather. We can't predict whether it will rain one month from today, but we do know the rules that govern the rain, even though we can't always calculate their consequences. I see nothing about the human mind any more than about the weather that stands out as beyond the hope of understanding as a consequence of impersonal laws acting over billions of years.

There do not seem to be any exceptions to this natural order, any miracles. I have the impression that these days most theologians are embarrassed by talk of miracles, but the great monotheistic faiths are founded on miracle stories—the burning bush, the empty tomb, an angel dictating the Koran to Mohammed—and some of these faiths teach that miracles continue at the present day. The evidence for all these miracles seems to me to be considerably weaker than the evidence for cold fusion, and I don't believe in cold fusion. Above all, today we understand that even human beings are the result of natural selection acting over millions of years of breeding and eating.

I'd guess that if we were to see the hand of the designer anywhere, it would be in the fundamental principles, the final laws of nature, the book of rules that govern all natural phenomena. We don't know the final laws yet, but as far as we have been able to see, they are utterly impersonal and quite without any special role for life. There is no life force. As Richard Feynman has said, when you look at the universe and understand its laws, "the theory that it is all arranged as a stage for God to watch man's struggle for good and evil seems inadequate."

True, when quantum mechanics was new, some physicists thought that it put humans back into the picture, because the principles of quantum mechanics tell us how to calculate the probabilities of various results that might be found by a human observer. But, starting with the work of Hugh Everett forty years ago, the tendency of physicists who think deeply about these things has been to reformulate quantum mechanics in an entirely objective way, with observers treated just like everything else. I don't know if this program has been completely successful yet, but I think it will be.

I have to admit that, even when physicists will have gone as far as they can go, when we have a final theory, we will not have a completely satisfying picture of the world, because we will still be left with the question "Why?" Why this theory, rather than some other theory? For example, why is the world described by quantum mechanics? Quantum mechanics is the one part of our present physics that is likely to survive intact in any future theory, but there is nothing logically inevitable about quantum mechanics; I can imagine a universe governed by Newtonian mechanics instead. So there seems to be an irreducible mystery that science will not eliminate.

But religious theories of design have the same problem. Either you mean something definite by a God, a designer, or you don't. If you don't, then what are we talking about? If you do mean something definite by "God" or "design," if for instance you believe in a God who is jealous, or loving, or intelligent, or whimsical, then you still must confront the question "Why?" A religion may assert that the universe is governed by that sort of God, rather than some other sort of God, and it may offer evidence for this belief, but it cannot explain why this should be so.

In this respect, it seems to me that physics is in a better position to give us a partly satisfying explanation of the world than religion can ever be, because although physicists won't be able to explain why the laws of nature are what they are and not something completely different, at least we may be able to explain why they are not slightly different. For instance, no one has been able to think of a logically consistent alternative to quantum mechanics that is only slightly different. Once you start trying to make small changes in quantum mechanics, you get into theories with negative probabilities or other logical absurdities. When you combine quantum mechanics with relativity you increase its logical fragility. You find that unless you arrange the theory in just the right way you get nonsense, like effects preceding causes, or infinite probabilities. Religious theories, on the other hand, seem to be infinitely flexible, with nothing to prevent the invention of deities of any conceivable sort.

Now, it doesn't settle the matter for me to say that we cannot see the hand of a designer in what we know about the fundamental principles of science. It might be that, although these principles do not refer explicitly to life, much less human life, they are nevertheless craftily designed to bring it about.

Some physicists have argued that certain constants of nature have values that seem to have been mysteriously fine-tuned to just the values that allow for the possibility of life, in a way that could only be explained by the intervention of a designer with some special concern for life. I am not impressed with these supposed instances of fine-tuning. For instance, one of the most frequently quoted examples of fine-tuning has to do with a property of the nucleus of the carbon atom. The matter left over from the first few minutes of the universe was almost entirely hydrogen and helium, with virtually none of the heavier elements like carbon, nitrogen, and oxygen that seem to be necessary for life. The heavy elements that we find on Earth were built up hundreds of millions of years later in a first generation of stars, and then spewed out into the interstellar gas out of which our solar system eventually formed.

The first step in the sequence of nuclear reactions that created the heavy elements in early stars is usually the formation of a carbon nucleus out of three helium nuclei. There is a negligible chance of producing a carbon nucleus in its normal state (the state of lowest energy) in collisions of three helium nuclei, but it would be possible to produce appreciable amounts of carbon in stars if the carbon nucleus could exist in a radioactive state with an energy roughly 7 million electron volts (MeV) above the energy of the normal state, matching the energy of three helium nuclei, but (for reasons I'll come to presently) not more than 7.7 MeV above the normal state.

This radioactive state of a carbon nucleus could be easily formed in stars from three helium nuclei. After that, there would be no problem in producing ordinary carbon; the carbon nucleus in its radioactive state would spontaneously emit light and turn into carbon in its normal nonradioactive state, the state found on Earth. The critical point in producing carbon is the existence of a radioactive state that can be produced in collisions of three helium nuclei.

In fact, the carbon nucleus is known experimentally to have just such a radioactive state, with an energy 7.65 MeV above the normal state. At first sight this may seem like a pretty close call; the energy of this radioactive state of carbon misses being too high to allow the formation of carbon (and hence of us) by only 0.05 MeV, which is less than 1 percent of 7.65 MeV. It may appear that the constants of nature on which the properties of all nuclei depend have been carefully fine-tuned to make life possible.

Looked at more closely, the fine-tuning of the constants of nature here does not seem so fine. We have to consider the reason why the formation of carbon in stars requires the existence of a radioactive state of carbon with an energy not more than 7.7 MeV above the energy of the normal state. The reason is that the carbon nuclei in this state are actually formed in a two-step process: first, two helium nuclei combine to form the unstable nucleus of a beryllium isotope, beryllium 8, which occasionally, before it falls apart, captures another helium nucleus, forming a carbon nucleus in its radioactive state, which then decays into normal carbon. The total energy of the beryllium 8 nucleus and a helium nucleus at rest is 7.4 MeV above the energy of the normal state of the carbon nucleus; so if the energy of the radioactive state of carbon were more than 7.7 MeV it could only be formed in a collision of a helium nucleus and a beryllium 8 nucleus if the energy of motion of these two nuclei were at least 0.3 MeV—an energy which is extremely unlikely at the temperatures found in stars.

Thus the crucial thing that affects the production of carbon in stars is not the 7.65 MeV energy of the radioactive state of carbon above its normal state, but the 0.25 MeV energy of the radioactive state, an unstable composite of a beryllium 8 nucleus and a helium nucleus, above the energy of those nuclei at rest.[1] This energy misses being too high for the production of carbon by a fractional amount of 0.05 MeV/0.25 MeV, or 20 percent, which is not such a close call after all.

This conclusion about the lessons to be learned from carbon synthesis is somewhat controversial. In any case, there is one constant whose value does seem remarkably well adjusted in our favor. It is the energy density of empty space, also known as the cosmological constant. It could have any value, but from first principles one would guess that this constant should be very large, and could be positive or negative. If large and positive, the cosmological constant would act as a repulsive force that increases with distance, a force that would prevent matter from clumping together in the early universe, the process that was the first step in forming galaxies and stars and planets and people. If large and negative the cosmological constant would act as an attractive force increasing with distance, a force that would almost immediately reverse the expansion of the universe and cause it to recollapse, leaving no time for the evolution of life. In fact, astronomical observations show that the cosmological constant is quite small, very much smaller than would have been guessed from first principles.

It is still too early to tell whether there is some fundamental principle that can explain why the cosmological constant must be this small. But even if there is no such principle, recent developments in cosmology offer the possibility of an explanation of why the measured values of the cosmological constant and other physical constants are favorable for the appearance of intelligent life. According to the "chaotic inflation" theories of André Linde and others, the expanding cloud of billions of galaxies that we call the Big Bang may be just one fragment of a much larger universe in which Big Bangs go off all the time, each one with different values for the fundamental constants.

In any such picture, in which the universe contains many parts with different values for what we call the constants of nature, there would be no difficulty in understanding why these constants take values favorable to intelligent life. There would be a vast number of Big Bangs in which the constants of nature take values unfavorable for life, and many fewer where life is possible. You don't have to invoke a benevolent designer to explain why we are in one of the parts of the universe where life is possible: in all the other parts of the universe there is no one to raise the question.[2] If any theory of this general type turns out to be correct, then to conclude that the constants of nature have been fine-tuned by a benevolent designer would be like saying, "Isn't it wonderful that God put us here on Earth, where there's water and air and the surface gravity and temperature are so comfortable, rather than some horrid place, like Mercury or Pluto?" Where else in the solar system other than on Earth could we have evolved?

Reasoning like this is called "anthropic." Sometimes it just amounts to an assertion that the laws of nature are what they are so that we can exist, without further explanation. This seems to me to be little more than mystical mumbo jumbo. On the other hand, if there really is a large number of worlds in which some constants take different values, then the anthropic explanation of why in our world they take values favorable for life is just common sense, like explaining why we live on Earth rather than Mercury or Pluto. The actual value of the cosmological constant, recently measured by observations of the motion of distant supernovas, is about what you would expect from this sort of argument: it is just about small enough so that it does not interfere much with the formation of galaxies. But we don't yet know enough about physics to tell whether there are different parts of the universe in which what are usually called the constants of physics really do take different values. This is not a hopeless question; we will be able to answer it when we know more about the quantum theory of gravitation than we do now.

It would be evidence for a benevolent designer if life were better than could be expected on other grounds. To judge this, we should keep in mind that a certain capacity for pleasure would readily have evolved through natural selection, as an incentive to animals who need to eat and breed in order to pass on their genes. It may not be likely that natural selection on any one planet would produce animals who are fortunate enough to have the leisure and the ability to do science and think abstractly, but our sample of what is produced by evolution is very biased, by the fact that it is only in these fortunate cases that there is anyone thinking about cosmic design. Astronomers call this a selection effect.

The universe is very large, and perhaps infinite, so it should be no surprise that, among the enormous number of planets that may support only unintelligent life and the still vaster number that cannot support life at all, there is some tiny fraction on which there are living beings who are capable of thinking about the universe, as we are doing here. A journalist who has been assigned to interview lottery winners may come to feel that some special providence has been at work on their behalf, but he should keep in mind the much larger number of lottery players whom he is not interviewing because they haven't won anything. Thus, to judge whether our lives show evidence for a benevolent designer, we have not only to ask whether life is better than would be expected in any case from what we know about natural selection, but we need also to take into account the bias introduced by the fact that it is we who are thinking about the problem.

This is a question that you all will have to answer for yourselves. Being a physicist is no help with questions like this, so I have to speak from my own experience. My life has been remarkably happy, perhaps in the upper 99.99 percentile of human happiness, but even so, I have seen a mother die painfully of cancer, a father's personality destroyed by Alzheimer's disease, and scores of second and third cousins murdered in the Holocaust. Signs of a benevolent designer are pretty well hidden.

The prevalence of evil and misery has always bothered those who believe in a benevolent and omnipotent God. Sometimes God is excused by pointing to the need for free will. Milton gives God this argument in *Paradise Lost*:

> I formed them free, and free they must remain
> Till they enthral themselves: I else must change
> Their nature, and revoke the high decree
> Unchangeable, eternal, which ordained
> Their freedom; they themselves ordained their fall.

It seems a bit unfair to my relatives to be murdered in order to provide an opportunity for free will for Germans, but even putting that aside, how does free will account for cancer? Is it an opportunity of free will for tumors?

I don't need to argue here that the evil in the world proves that the universe is not designed, but only that there are no signs of benevolence that might have shown the hand of a designer. But in fact the perception that God cannot be benevolent is very old. Plays by Aeschylus and Euripides make a quite explicit statement that the gods are selfish and cruel, though they expect better behavior from humans. God in the Old Testament tells us to bash the heads of infidels and demands of us that we be willing to sacrifice our children's lives at his orders, and the God of traditional Christianity and Islam damns us for eternity if we do not worship him in the right manner. Is this a nice way to behave? I know, I know, we are not supposed to judge God according to human standards, but you see the problem here: If we are not yet convinced of his existence, and are looking for signs of his benevolence, then what other standards can we use?

The issues that I have been asked to address here will seem to many to be terribly old-fashioned. The "argument from design" made by the English theologian William Paley is not on most people's minds these days. The prestige of religion seems today to derive from what people take to be its moral influence, rather than from what they may think has been its success in accounting for what we see in nature. Conversely, I have to admit that, although I really don't believe in a cosmic designer, the reason that I am taking the trouble to argue about it is that I think that on balance the moral influence of religion has been awful.

This is much too big a question to be settled here. On one side, I could point out endless examples of the harm done by religious enthusiasm, through a long history of pogroms, crusades, and jihads. In our own century it was a

Muslim zealot who killed Sadat, a Jewish zealot who killed Rabin, and a Hindu zealot who killed Gandhi. No one would say that Hitler was a Christian zealot, but it is hard to imagine Nazism taking the form it did without the foundation provided by centuries of Christian anti-Semitism. On the other side, many admirers of religion would set countless examples of the good done by religion. For instance, in his recent book *Imagined Worlds*, the distinguished physicist Freeman Dyson has emphasized the role of religious belief in the suppression of slavery. I'd like to comment briefly on this point, not to try to prove anything with one example but just to illustrate what I think about the moral influence of religion.

It is certainly true that the campaign against slavery and the slave trade was greatly strengthened by devout Christians, including the Evangelical layman William Wilberforce in England and the Unitarian minister William Ellery Channing in America. But Christianity, like other great world religions, lived comfortably with slavery for many centuries, and slavery was endorsed in the New Testament. So what was different for antislavery Christians like Wilberforce and Channing? There had been no discovery of new sacred scriptures, and neither Wilberforce nor Channing claimed to have received any supernatural revelations. Rather, the eighteenth century had seen a widespread increase in rationality and humanitarianism that led others—for instance, Adam Smith, Jeremy Bentham, and Richard Brinsley Sheridan—also to oppose slavery, on grounds having nothing to do with religion. Lord Mansfield, the author of the decision in Somersett's Case, which ended slavery in England (though not its colonies), was no more than conventionally religious, and his decision did not mention religious arguments. Although Wilberforce was the instigator of the campaign against the slave trade in the 1790s, this movement had essential support from many in Parliament like Fox and Pitt, who were not known for their piety. As far as I can tell, the moral tone of religion benefited more from the spirit of the times than the spirit of the times benefited from religion.

Where religion did make a difference, it was more in support of slavery than in opposition to it. Arguments from scripture were used in Parliament to defend the slave trade. Frederick Douglass told in his *Narrative* how his condition as a slave became worse when his master underwent a religious conversion that allowed him to justify slavery as the punishment of the children of Ham. Mark Twain described his mother as a genuinely good person, whose soft heart pitied even Satan, but who had no doubt about the legitimacy of slavery, because in years of living in antebellum Missouri she had never heard any sermon opposing slavery, but only countless sermons preaching that slavery was God's will. With or without religion, good people can behave well and bad people can do evil; but for good people to do evil—that takes religion.

In an e-mail message from the American Association for the Advancement of Science I learned that the aim of this conference is to have a constructive dialogue between science and religion. I am all in favor of a dialogue between science and religion, but not a constructive dialogue. One of the great achievements of science has been, if not to make it impossible for intelligent people to be religious, then at least to make it possible for them not to be religious. We should not retreat from this accomplishment.

Notes

1. This was pointed out in a 1989 paper by M. Livio, D. Hollowell, A. Weiss, and J. W. Truran ("The Anthropic Significance of the Existence of an Excited State of 12C," *Nature* 340, no. 6231 [July 27, 1989]). They did the calculation quoted here of the 7.7 MeV maximum energy of the radioactive state of carbon, above which little carbon is formed in stars.

2. The same conclusion may be reached in a more subtle way when quantum mechanics is applied to the whole universe. Through a reinterpretation of earlier work by Stephen Hawking, Sidney Coleman has shown how quantum mechanical effects can lead to a split of the history of the universe (more precisely, in what is called the wave function of the universe) into a huge number of separate possibilities, each one corresponding to a different set of fundamental constants. See Sidney Coleman, "Black Holes as Red Herrings: Topological Fluctuations and the Loss of Quantum Coherence," *Nuclear Physics* B307 (1988): 867.

4

ANTHROPIC DESIGN
Does the Cosmos Show Evidence of Purpose?

VICTOR J. STENGER

Poking Out of the Noise

For about a decade now, an increasing number of scientists and theologians have been asserting, in popular articles and books, that they can detect a signal of cosmic purpose poking its head out of the noisy data of physics and cosmology (see, for example, Swinburne 1990; Ellis 1993; Ross 1995). This claim has been widely reported in the media (see, for example, Begley 1998; Easterbrook 1998), perhaps misleading laypeople into thinking that some kind of new scientific consensus is developing in support of supernatural beliefs. In fact, none of this purported evidence can be found in the pages of scientific journals, which continue to operate within a framework in which all physical phenomena are assumed natural. As the argument goes, the data are said to reveal a universe that is exquisitely fine-tuned for the production of life. This precise balancing act is claimed to be too unlikely to be the result of mindless

Victor J. Stenger's "Anthropic Design: Does the Cosmos Show Evidence of Purpose?" appeared in the *Skeptical Inquirer* 23, no. 4 (July/August 1999). It is a much abridged version of "The Anthropic Coincidences: A Natural Explanation," which appeared in the British *Skeptical Intelligencer* 3, no. 3 (July 1999).

chance. An intelligent, purposeful, and indeed personal Creator must have made things the way they are.

As cosmologist and Quaker George Ellis explains it: "The symmetries and delicate balances we observe require an extraordinary coherence of conditions and cooperation of laws and effects, suggesting that in some sense they have been purposefully designed" (Ellis 1993, 97). Others have been less restrained in insisting that God is now *required* by the data and that this God must be the God of the Christian Bible (see, for example, Ross 1995).

The fine-tuning argument is based on the fact that life on Earth is very sensitive to the values of several fundamental physical constants. The tiniest change in any of these would result in changes so drastic that life as we know it would not exist. The delicate connections between physical constants and life are called the *anthropic coincidences* (Carter 1974; Barrow and Tipler 1986). The name is a misnomer. Human life is not singled out in any special way. At most, the coincidences show that the production of carbon and the other elements that make Earth's life possible required a sensitive balance of physical parameters.

For example, if the gravitational attraction between protons in stars had not been many orders of magnitude weaker than their electrical repulsion, stars would have collapsed long before nuclear processes could build up the chemical periodic table from the original hydrogen and deuterium. Furthermore, the element-synthesizing reactions in stars depend sensitively on the properties and abundances of deuterium and helium produced in the early universe. Deuterium would not exist if the neutron-proton mass difference were just slightly displaced from its actual value; neutrons, unstable in a free state, were stored in deuterium for their later use in building the elements.

The existing relative abundances of hydrogen and helium also implies a close balance of the relative strengths of the gravitational and weak nuclear forces. Given a slightly stronger weak force, the universe would be 100 percent hydrogen as all neutrons decayed away before assembling into deuterium and helium. A slightly weaker weak force and we would have a universe that is 100 percent helium; in that case neutrons would not have decayed and left the excess of protons that formed hydrogen. Neither of these extremes would have allowed for the existence of stars and life, as we know it, based on carbon chemistry. Barrow and Tipler (1986) list many other such "coincidences," some remarkable, others somewhat strained.

Interpreting the Coincidences

The interpretation of the anthropic coincidences in terms of purposeful design should be recognized as yet another variant of the ancient *argument from design* that has appeared in many different forms over the ages. The anthropic design argument asks: How can the universe possibly have obtained the unique set of physical constants it has, so exquisitely fine-tuned for life as they are, except by purposeful design—design with life and perhaps humanity in mind?

This argument, however, has at least one fatal flaw. It makes the wholly unwarranted assumption that *only one type of life is possible*—the particular form of carbon-based life we have here on Earth. Even if this is an unlikely result of chance, some form of life could still be a likely result. It is like arguing that a particular card hand is so improbable that it must have been preordained.

Based on recent studies in the sciences of complexity and "artificial life" computer simulations, sufficient complexity and long life appear to be primary conditions for a universe to contain some form of reproducing, evolving structures. This can happen with a wide range of physical parameters, as has been demonstrated (Stenger 1995). The fine-tuners have no basis in current knowledge for assuming that life is impossible except for a very narrow, improbable range of parameters.

Amusingly, the new cosmic creationists contradict the traditional design argument of the biological creationists, that the universe is so *uncongenial* to life that life could not have evolved naturally. The new creationists now tell us that the universe is so *congenial* to life that the universe could not have evolved naturally.

Since all scientific explanations until now have been natural, then it would seem that the first step, before asserting purposeful design, is to seek a natural explanation for the anthropic coincidences. Such a quest would avoid the invocation of supernatural agency until it is absolutely required by the data.

The Natural Scenario

For almost two decades, the *inflationary big bang* has been the standard model of cosmology (Guth 1981, 1997; Linde 1987, 1990, 1994). We keep hearing, again from the unreliable popular media, that the big bang theory is in trouble and the inflationary model is dead. In fact, no viable substitute has been proposed that has nearly the equivalent explanatory power.

The inflationary big bang offers a plausible, natural scenario for the uncaused origin and evolution of the universe, including the formation of order and structure—without the violation of any laws of physics. These laws themselves are now understood far more deeply than before, and we are beginning to grasp how they too could have come about naturally. The natural scenario I will describe here has not yet risen to the exalted status of a scientific theory. However, the fact that it is consistent with all current knowledge and cannot be ruled out at this time demonstrates that no rational basis exists for introducing the added hypothesis of supernatural creation. Such a hypothesis is simply not required by the data.

According to the proposed natural scenario, by means of a random quantum fluctuation the universe "tunneled" from pure vacuum ("nothing") to what is called a *false vacuum*, a region of space that contains no matter or radiation but is not quite nothing. The space inside a bubble of false vacuum is curved, or warped, and a small amount of energy is stored in that curvature, like the potential energy of a strung bow. This ostensible violation of energy conservation is allowed by the Heisenberg uncertainty principle for sufficiently small time intervals.

The bubble then inflated exponentially and the universe grew by many orders of magnitude in a tiny fraction of a second. (For a not-too-technical discussion and original references, see Stenger 1990.) As the bubble expanded, its curvature energy transformed (naturally) into matter and radiation. Inflation stopped, and the more linear big bang expansion we now experience commenced. As the universe cooled, its structure spontaneously froze out—just as formless water vapor freezes into snowflakes whose unique and complex patterns arise from a combination of symmetry and randomness.

In our universe, the first galaxies began to assemble after about a billion years, eventually evolving into stable systems where stars could live out their lives and populate the interstellar medium with the complex chemical elements, such as carbon, needed for the formation of life.

So how did our universe happen to be so "fine-tuned" as to produce wonderful, self-important carbon structures? As I explained above, we have no reason to assume that ours is the only possible form of life. Life of some sort could have happened regardless of the form the universe took—however the crystals on the arm of the snowflake happened to get arranged by chance.

If we have no reason to assume ours is the only life-form, we also have no reason to assume that ours is the only universe. Many universes can exist, with all possible combinations of physical laws and constants. In that case, we just happen to be in the particular one that was suited for the evolution of our form of life. When cosmologists refer to the *anthropic principle*, this is all they usually mean. Since we live in this universe, we can assume it possesses qualities suitable for our existence. Humans evolved eyes sensitive to the region of electromagnetic spectrum from red to violet because the atmosphere is transparent in that range. Yet some would have us think that the causal action was the opposite, that the atmosphere of Earth was designed to be transparent from red to violet because human eyes are sensitive in that range. Stronger versions of the anthropic principle, which assert that the universe is somehow actually *required* to produce intelligent "information-processing systems" (Barrow and Tipler 1986), are not taken seriously by most scientists or philosophers.

The existence of many universes is consistent with all we know about physics and cosmology (Smith 1990; Smolin 1992, 1997; Linde 1994; Tegmark 1997). Some theologians and scientists dismiss the notion as a gross violation of Occam's razor (see, for example, Swinburne 1990). It is not. No new hypothesis is needed to consider multiple universes. In fact, it takes an added hypothesis to rule them out—a super law of nature that says only one universe can exist. But we know of no such law, so we would violate Occam's razor to insist on only one universe. Another way to express this is with lines from T. H. White's *The Once and Future King:* "Everything not forbidden is compulsory."

The hundred billion galaxies of our visible universe, each with a hundred billion stars, is but a grain of sand on the Sahara that exists beyond our horizon, grown out of that single, original bubble of false vacuum. An endless number of such bubbles can very well exist, each itself nothing but a grain of sand on the Sahara of all existence. On such a Sahara, nothing is too improbable to have happened by chance.

Acknowledgments

I have greatly benefitted from discussions on this subject with Ricardo Aler Mur, Samantha Atkins, John Chalmers, Scott Dalton, Keith Douglas, Ron Ebert, Simon Ewins, Jim Humphreys, Bill Jefferys, Kenneth Porter, Wayne Spencer, Quentin Smith, and Ed Weinmann.

References

Barrow, John D., and Frank J. Tipler. 1986. *The Anthropic Cosmological Principle.* Oxford: Oxford University Press.

Begley, Sharon. 1998. "Science Finds God." *Newsweek,* July 20, 46.

Carter, Brandon. 1974. "Large Number Coincidences and the Anthropic Principle in Cosmology." In *Confrontation of Cosmological Theory with Astronomical Data,* edited by M. S. Longair. Dordrecht: Reidel, 291–98. Reprinted in Leslie 1990.

Easterbrook, Greg. 1998. "Science Sees the Light." *New Republic,* October 12, 24–29.

Ellis, George. 1993. *Before the Beginning: Cosmology Explained.* London and New York: Boyars/Bowerdean.

Guth, Alan. 1981. "Inflationary Universe: A Possible Solution to the Horizon Flatness Problems." *Physical Review* D23, 347–56.

————. 1997. *The Inflationary Universe.* New York: Addison-Wesley.

Leslie, John. 1990. *Physical Cosmology and Philosophy.* New York: Macmillan.

Linde, Andre. 1987. "Particle Physics and Inflationary Cosmology." *Physics Today* 40, 61–68.

————. 1990. *Particle Physics and Inflationary Cosmology.* New York: Academic Press.

Linde, Andre. 1994. "The Self-reproducing Inflationary Universe." *Scientific American,* November, 48–55.

Ross, Hugh. 1995. *The Creator and the Cosmos: How the Greatest Scientific Discoveries of the Century Reveal God.* Colorado Springs: Navpress.

Smolin, Lee. 1992. "Did the Universe Evolve?" *Classical and Quantum Gravity* 9, 173–91.

————. 1997. *The Life of the Cosmos.* Oxford: Oxford University Press.

Smith, Quentin. 1990. "A Natural Explanation of the Existence and Laws of Our Universe." *Australasian Journal of Philosophy* 68, 22–43.

Stenger, Victor J. 1988. *Not by Design: The Origin of the Universe.* Amherst, N.Y.: Prometheus Books.

————. 1990. "The Universe: The Ultimate Free Lunch." *European Journal of Physics* 11, 236.

————. 1995. *The Unconscious Quantum: Metaphysics in Modern Physics and Cosmology.* Amherst, N.Y.: Prometheus Books.

Swinburne, Richard. 1990. "Argument from the Fine-Tuning of the Universe." *In* Leslie 1990, 154–73.

Tegmark, Max. 1997. "Is 'the Theory of Everything' Merely the Ultimate Ensemble Theory?" To be published in *Annals of Physics.*

5

FROM THE ANTHROPIC PRINCIPLE TO THE SUPERNATURAL

DAVID A. SHOTWELL

Thomas Aquinas was not content to base religion upon faith, but gave purported proofs of God's existence. These have been elaborated upon by theologians and metaphysicians, and a number of apologists have attempted to enlist science itself in support of supernaturalism. An example is provided by Patrick Glynn in his recent book *God: The Evidence*.[1] I will comment upon some of his claims but leave to others the analysis of his appeal to "near-death experiences" as evidence for immortality.

Glynn is very confident. Five hundred years after the birth of Copernicus, he says, we have witnessed the philosophical overthrow of the Copernican revolution: Although scientists assumed naïvely that humans are not the favorites of the universe and that it does not exist for our benefit, they now know, or should know, better. The initial step toward that conclusion was taken by the cosmologist Brandon Carter, who introduced his "anthropic principle" in 1973. It is summarized by Glynn in the statement that the laws of nature are precisely those needed to provide a universe capable of producing life. Because life does exist, this supposedly explains why the laws and constants

---

David A. Shotwell's "From the Anthropic Principle to the Supernatural" originally appeared in the *Skeptical Inquirer* 22, no. 6 (November/December 1998).

of physics are as they are. If they were not, we would not be here; but we are here. This is said to be a bold new type of explanation. Glynn uses it as a first step toward a leap of faith, upon which he elaborates at length: The laws of physics are the work of an intelligent Creator, whose objective was to produce human beings.

Why human beings? It is easy to think of alternatives. Perhaps the Creator, if there is one, really wanted to produce dinosaurs, and we are an unimportant by-product of the enterprise. It might be objected that dinosaurs are, after all, extinct. That is true enough, but does anyone think that we will be here forever? If we last one tenth as long as did the dinosaurs it will be an impressive achievement.

I do not insist upon dinosaurs. Instead, I would put my money on insects, the most successful of all life-forms. The total number of them is somewhere in the trillions, as compared with only a few billion humans. The entomologic principle, newly formulated by me, states that the Creator had insects in mind all along, and fine-tuned his laws to produce them. Why, you may ask, would he want to create insects? I am sure that I don't know. But I also don't know why he would want to create humans.

The concept of randomness is misinterpreted by Glynn. He refers to the "random universe cosmology" that underlies atheistic philosophies, attributes to scientists the view that randomness engenders order, and says that you cannot explain away the order in nature by reference to a purely random process. These statements make it clear that for Glynn a random event or process is simply one that is not produced by an intelligent agent. His argument therefore reduces to the question-begging assertion that orderly structures and processes must be the work of such an agent because they could not occur otherwise. That is the essence of the traditional argument to design, which does not depend upon the anthropic principle.

Scientists do not explain, or explain away, the order of nature by invoking pure randomness. With the possible exception of submicroscopic processes subject to quantum mechanical indeterminacy, they believe instead that events occur in accordance with natural laws. When a liquid solidifies to form crystals, the crystals have an orderly structure, and neither the structure nor the process is a random one; it is a result of the laws of physics. An unlimited number of other examples could be given. Glynn would reply that the laws themselves have not been explained. In many cases, however, they have been; the reduction of thermodynamics to statistical mechanics is the standard example. This, of course, is not what he is demanding. Science, he says, must surrender its pretension to answer the ultimate questions, and reason is an imperfect guide to ultimate truths. He implies, therefore, that when reason has taken us part of the way we should abandon it and rely upon unreason. It will lead us on to God, whose creative activities provide the answers to ultimate questions, in particular those concerning the existence of the universe and its laws.

I have a rival hypothesis to propose. Let us assume that each subatomic particle is inhabited by a ghostly little gremlin. Each gremlin maintains the existence of its particle by a continuous creative act and is in instantaneous telepathic communication with all of the others. By this means they cooperate to produce the universe and its lawful behavior. This hypothesis "explains" everything that exists and every event that occurs. It accounts for those nasty features of the world that are so troublesome to believers in an omnipotent and perfectly benevolent creator; they exist because the gremlins are mischievous and, in some respects, malevolent.

I advance this "theory" not to make a joke, but to make a point. If you admit the supernatural into your calculations, anything goes. That is why a supernatural explanation is useless to a scientist, however pious he may be on Sundays. It provides no direction for research, suggests no testable hypotheses, and gives no reason to expect one result rather than another from any observation or experiment.

Questions such as "Why does the universe exist?" and "Why does it follow one set of laws rather than another?" are regarded by Glynn as ultimate. Magic, personified in God, is his answer. But irreverent critics can reply by asking how an immaterial agent created matter, how he controls its behavior, and why he exists rather than nothing at all. The only answer we are given is, in effect, that the supernatural is beyond human comprehension. Mysteries, however, are not explanations, and it appears that only theologians and their allies are licensed to pretend that they are.

Note

1. Patrick Glynn, *God: The Evidence* (Rocklin, Calif.: Prima, 1999).

6

"GOD'S GOOF," AND THE UNIVERSE THAT KNEW WE WERE COMING

OWEN GINGERICH

At the end of October 2001, while I was in Prague helping to commemorate the death of Tycho Brahe exactly four centuries ago, I had a rather vivid dream in which I was addressing some unspecified listeners. Now in my many years of teaching "The Astronomical Perspective," a general education course at Harvard, I had had occasion to use Robert Frost's "Fire and Ice" poem in the cosmological context of the far future universe, as a metaphor for the differing eschatological scenarios, whether the universe would end in an icy whimper or a fiery Big Crunch, and I had been lamenting that since I have now retired from teaching, I would no longer have occasion to recite the poem. That's the background, but curiously enough, in my dream I found I was putting the poem to a different use.

I remarked to that dream-world audience that Frost had written the poem almost entirely in monosyllables, with half a dozen two-syllable words, and only one word of three syllables, at the climax of the poem:

Owen Gingerich's "'God's Goof,' and the Universe that Knew We Were Coming" was originally delivered at the conference "Science and Religion: Are They Compatible?" sponsored by the Center for Inquiry, and held in Atlanta, Georgia, in November 2001.

Some say the world will end in fire.
Some say in ice.
From what I've tasted of desire,
I hold with those who favor fire.
But if it had to perish twice,
I think I know enough of hate
To say that for destruction ice
Is also great
And would suffice.

At its heart this is a poem about evil, about the narrow line between passion or desire and hate and destruction. To appreciate evil is one of the hallmarks of humanity. The tree of good and evil is one of the earliest metaphors in the Hebrew scriptures. Its understanding becomes the origin of conscience, the delineation between man and beast.

In my dream I went on to mention a friend who I described as one of the most brilliant and articulate scientists of our age, Steven Weinberg. Two years ago I invited him and Sir John Polkinghorne to debate the issue of "Is the Universe Designed?" at an AAAS symposium at the Smithsonian in Washington. It was a memorable debate, one that may go down in lore along with the 1920 Shapley-Curtis debate on the scale of the universe, which was held in the very same auditorium at the National Museum of Natural History. Now I knew that Steve had rejected the idea of a Designer-Creator, for he had ended his best-selling book, *The First Three Minutes,* with the statement that "the more the universe seems comprehensible, the more it also seems pointless." But I was fascinated to learn in the course of the debate that he did not reject the existence of a Creator-God for any *scientific* reasons. Steve is smart enough to know that you can no more disprove the existence of God on scientific grounds than you can prove the existence of God from the evidence of science, *pace* Paley. In fact, Weinberg's rejection of a Designer God and a designed universe rests on his distress in coping with the existence of evil in our world, whether it is the Holocaust or the evolutionary struggle for survival or the starvation of children in Africa or Afghanistan.

It was at this point that my dream abruptly ended. I had neither solved the challenge of theodicy nor had I reviewed the equally cogent and eloquent response presented at the Washington debate by John Polkinghorne.[1] In any event, it is quite clear to me that there are alternative ways of making sense of our universe, and that some of us weigh the factors differently. For those of us who are theists, the powerful question, "Why is there something rather than nothing?" requires an answer in anthropocentric terms, compatible with the evidence that the universe seems peculiarly congenial to the emergence of self-conscious, reflective life, suggesting that, mysteriously, there is purpose and intention, and that we are part of that purpose and intention. In the words of Freeman Dyson, it's a universe that seems to have known that we were coming.

There is, within our universe, an astonishing confluence of parameters that are tuned just right to make a world habitable by intelligent life. Half a dozen of these dimensionless parameters are delineated in Sir Martin Rees' recent book, *Just Six Numbers*. The numbers the Astronomer Royal has chosen are, as far as we know, unrelated to each other, that is, each can be independently set, yet each must fall within a particular range to make a universe in which habitable environments can exist.

This morning I would like to mention yet another odd feature of our universe that turns out to be unexpectedly significant in allowing us to be here.

In 1913 the Harvard professor L. J. Henderson published a book entitled *The Fitness of the Environment* and there he extolled the remarkable life-enhancing properties of water as well as pointing out the unique properties of the carbon atom, including the fact that carbon can bond with itself in a vastly larger number of combinations than any other atom. It is this wonderful property that makes complex organic chemistry possible.

Of course, these unique properties would have been of little avail if it were not for the substantial abundance of oxygen and carbon. But since hydrogen and oxygen rank number one and three respectively in cosmic abundances, water is guaranteed to be ubiquitous throughout the universe, and carbon comes in as number four in cosmic abundance. However, neither oxygen nor carbon emerged in the first three minutes of the Big Bang. At first glance, this might be labeled "God's Goof." That's how the physicist George Gamow felt when he discovered the presumed flaw in the nature of the light elements that prevented the heavier elements from forming. In the first minute of the Big Bang, energetic photons transformed into protons, and these fused into deuterium (nuclear particles of mass 2), tritium (nuclear particles of mass 3), and alpha particles (which would serve as mass 4 nuclei of helium atoms). But there was no stable mass five, so at that point the fusion process stopped, well short of the 12 needed for carbon or the 16 for oxygen. Gamow, who had an impish wit, then wrote his own version of Gen. 1:[2]

In the beginning God created Radiation and Ylem (a mixture of protons and neutrons). And the Ylem was without shape or number, and the nucleons were rushing madly upon the face of the deep.

And God said: "Let there be mass two." And there was mass two. And God saw deuterium, and it was good.

And God said: "Let there be mass three." And there was mass three. And God saw tritium, and it was good.

And God continued to call numbers until He came to the transuranium elements. But when He looked back on his work, He saw that it was not good. In the excitement of counting, He had missed calling for mass five, and so, naturally, no heavier elements could have been formed.

God was very disappointed by that slip and wanted to contract the universe again and start everything from the beginning. But that would be much too simple. Instead, being Almighty, God decided to make heavy elements in the most impossible way.

And so God said: "Let there be Hoyle." And there was Hoyle. And God saw Hoyle and told him to make heavy elements in any way he pleased.

And so Hoyle decided to make heavy elements in stars, and to spread them around by means of supernova explosions. But in doing so, Hoyle had to follow the blueprint of abundances which God prepared earlier when He had planned to make the elements from Ylem.

Thus, with the help of God, Hoyle made all heavy elements in stars, but it was so complicated that neither Hoyle, nor God, nor anybody else can now figure out exactly how it was done.

But far from being a design flaw in our universe, the missing mass five seems essential to our existence. Suppose that mass five were stable. Then, with the overwhelming abundance of protons in the opening minutes of the universe, atom building could have taken place in mass steps of one, right up the nuclear ladder toward iron. This would have left no special abundance of carbon and oxygen, two essential building blocks of life.

That there is no stable mass five means that in actuality the element building in the stars takes place in a two-step process, first when the hydrogen is converted into helium, and then with an abundance of helium, a second process whereby helium is built up into atoms whose nuclei consist of integer numbers of helium nuclei. This includes oxygen and carbon, which, as I have indicated, are the cosmically most abundant atoms after hydrogen and helium. Without the missing mass five (as well as several other puzzling details in the structures of these lighter elements), not only would we probably not have the life-giving abundance of carbon and oxygen, but we wouldn't have the long, slow hydrogen burning of main sequence stars. It is of course that tedious process that gives the stable solar environment in which the evolutionary sequences can work out.

What at first glance appeared to be God's mistake turns out to be one of God's most ingenious triumphs. Certainly the way our universe works, the fact that it takes a very long time to generate the heavier elements, depends critically on the lack of a stable mass five. In the absence of a nuclear ladder with easy steps of one in mass, the ladder goes up in steps of four, so that the production of the various intervening heavier elements is a complicated matter. The late Fred Hoyle was a leading player in figuring out these processes, as alluded to in Gamow's parody. On the basis of the fact that life does exist, he predicted that there must be a special resonance level in the carbon nucleus to account for its high abundance, a prediction that was borne out and helped bring the Nobel Prize to Willy Fowler, whose experimental work showed that the resonance really did exist. I am told that Fred Hoyle said that nothing has shaken his atheism as much as this discovery. Though I talked with him from time to time, unfortunately I never had enough nerve to ask him if his atheism had really been shaken by finding this nuclear resonance structure. However, the answer came rather clearly in the November 1981 issue of the Cal Tech alumni magazine, where he wrote:

> Would you not say to yourself, "Some supercalculating intellect must have designed the properties of the carbon atom, otherwise the chance of my finding such an atom through the blind forces of nature would be utterly minuscule." Of course you would. . . . A commonsense interpretation of the facts suggests that a superintellect has monkeyed with physics, as well as with chemistry and biology, and that there are no blind forces worth speaking about in nature. The numbers one calculates from the facts seem to me so overwhelming as to put this conclusion almost beyond question.[3]

These curious details of nuclear structure, along with the six numbers of Martin Rees' book, are among the many aspects of our universe that make it remarkably fertile for the existence of intelligent life—so much so as to cry out for some explanation. As Joel Primack and Nancy Abrams have written, "The logical possibilities are (a) that God designed the universe with us in mind; (b) that there are many universes, as suggested by the idea of cosmic inflation, and we naturally live in one suited for our sort of life; or (c) that only one set of values for the physical parameters is mathematically possible, which might in turn follow from a fundamental 'theory of everything.' Polkinghorne has argued that (a) is the most economical assumption. But physicists generally try to answer physical questions with physical theories, so naturally most physicists are working on (b) and (c)."[4]

Clearly these options have led us directly to the topic of this session, "New Cosmologies and Religion," for inflation and multiple universes are definitely part of new cosmologies. Indeed, it is option (b), the multiverse hypothesis that Martin Rees adopts in his books *Before the Beginning* and *Just Six Numbers.* In considering his six anthropic numbers, he has come up with an alternative way of looking at them. It is imaginable that the process that brought forth our universe—a fluctuation in a quantum vacuum, to use the jargon that simply cloaks a mystery beneath a name—that process could have brought forth uncountable additional universes, each with its own assortment of physical constants, even some with other than three dimensions. Andre Linde, one of the architects of the inflationary universe cosmology, has been an enthusiastic advocate of just such a cosmos, with one universe after another budding off from existing universes, and Martin Rees has suggested the name "multiverse" for this collection of universes. Given myriad universes, with a variety of properties, then naturally the intelligent, self-contemplating beings we call *Homo sapiens* would be found in precisely the one where the roulette brought up the numbers congenial to complex life-forms!

But it is not clear to me how distinct options (a) and (b) really are. Let me remind you that alternative universes would be in their own spaces, not in shared space. Not only could these alternative universes never bump into each other, but communication with, and therefore any physical evidence for, external universes would be forever lacking. While the mathematics might show that alternative universes *could* indeed exist, it would be a matter of faith that they *actually do* exist, with no hope of observational confirmation. Thus it would be rather difficult to distinguish between the logical status of option (a), that God designed our universe with the ultimate emergence of self-reflective consciousness as its intention, and (b) that in some mysterious and equally ununderstandable way there are myriad universes and we're in the one that, like the little bear's porridge, is just right. In effect, option (a) is a final cause, which was especially popular as an explanation in Aristotelian times and has more recently gone out of fashion, whereas option (b) is an equivalent efficient cause—maybe, though it hardly seems efficient to have so many universes!

Some cosmologists are deeply offended that anyone would even seriously entertain the notion of universes that we could never hope to detect. These cosmologists are the strict empiricists, but not the philosophers who often find that understanding the universe requires leaps of insight that go beyond "Just the facts, ma'am, just the facts." Actually all of us astronomers believe in the existence of unobservable parts of our own universe: that is, the past and the future, which we cannot observe but whose existence we deduce from the present. As we look out into space, we are looking backward in time, and we do not see the undefinable present at a distance. Rather, we are seeing the past. Consider a star a hundred light-years away. In a century from now our successors will observe today's present on that star—but it is an assumption that the star we now see as it was a century ago is still there and will be seen at the beginning of the twenty-second century.

Now let me remind you that Christians have long envisioned a world with which they have no physical contact, not the heavens, but Heaven, the empyrean. It is a totally other place, without evil and suffering, and where the inhabitants never grow old. It thus cannot be our present world remodeled, for the remodeling would strike at the very heart of all our physical understanding. To suspend the rules of our cosmos would be tantamount to being in another universe. Yet to deny the existence of such a universe, of anything other than our observed physical cosmos, would be to deny a traditional Christian doctrine.

So, I would like to explore how modern cosmological ideas can intersect with the ancient Christian ideas of a new heavens and a new earth, that is to say, a rational envisioning of a new and different physics, something that can no more be dismissed out of hand than the notion of a multiverse. And I say this knowing full well that there are many who in fact dismiss the multiverse concept as "mere metaphysics."

Traditionally the cosmos was seen on a comparatively human scale of both space and time. The biblical account described a universe created a very

long time ago in terms of human generations, but trivially short by any modern cosmological standards. Similarly, the starry sphere was seen as not all that
far above the earth. The levels of paradise described so vividly by Dante came
immediately beyond the firmament of stars. But note the difference in the
sources for the short time scale versus the short distance scale. The short time
scale appears at first glance to be anchored in the scriptures, whereas the place
of heaven, that is, the short distance scale, is a traditional interpretation, one
that finds its depiction in literature and art rather than in the Bible.
Nevertheless, the New Testament account of how Jesus came into the world
and how he ascended from it is deeply rooted in an antique biology and an
antique distance scale that stands in stark contrast to our modern scientific
understanding.

Unlike the ancient cosmos, the scale of today's theater for God's action is
mind-numbing in its sweep. There is no way to grasp the enormity of scale of
time and space except by making stepwise comparisons or some kind of analogies. When Christiaan Huygens gauged the distance to a nearby star Sirius, as
28,000 times farther away than the Sun, he recorded the enormity of that distance as follows:

> For if 25 years are required for a bullet out of a Cannon, with its ut-
> most Swiftness, to travel from the Sun to us; then by multiplying the
> number 27664 into 25, we shall find that such a Bullet would spend
> almost seven hundred thousand years in its Journey between us and the
> nearest of the fix'd Stars. And yet when in a clear night we look upon
> them, we cannot think them above some few miles over our heads.
> What I have here enquir'd into, is concerning the nearest of them. And
> what a prodigious number must there be besides of those which are
> placed so deep in the vast spaces of Heaven, as to be as remote from
> these as these are from the Sun! For if with our bare Eye we can observe
> above a thousand, and with a Telescope can discover ten or twenty
> times as many; what bounds of number must we set to those which are
> out of reach even of these Assistances! especially if we consider the infi-
> nite Power of God. Really, when I have bin reflecting thus with my self,
> methoughts all our Arithmetick was nothing, and we are vers'd but in
> the very Rudiments of Numbers, in comparison of this great Sum.[5]

Huygens's feeling is easily shared today, but he was probing only the fringe of
the vastness. He was one of the first to assign a reasonably correct velocity to

light itself, but it didn't occur to him to measure the distance to Sirius in terms of the flight time of light rather than the cannon ball. Today we put Sirius's distance at eight light-years, and eight years is easily within human grasp. But the diameter of the Milky Way Galaxy—100,000 light-years—exceeds ordinary human comprehension, a time interval vastly greater than all of human history. To go a thousand times farther takes us to distant clusters of galaxies, but that would delineate a sphere whose volume is still less than one ten-thousandth of a percent of the volume of the observable universe.

Indeed, we are mere specks in an abyss of space and time. God may see the little sparrow fall, but can we take seriously the concept of a caring Creator whose universe has somehow allowed us to emerge in this tiny backwater of the cosmos, overwhelmed by a sea of space and time? We have eaten of the tree of knowledge and see that we are naked.

Philosophically it hardly matters whether the universe is 100 million light-years in diameter or 14 billion—once we have left a human scale behind, it leaves us just as lost and insignificant whether we are talking millions or billions. But scientifically it matters a great deal whether we are in a universe 100 million years old or one 10 to 20 billion years old. A hundred million years puts us back to the age of the dinosaurs, but if the universe were only 100 million years old, neither the dinosaurs nor humankind would be here. Modern science shows us a universe in which the atoms required for life did not emerge in the immediate aftermath of the Big Bang. Rather, they simmered very slowly in the fiery furnaces of stellar interiors, gradually building up the abundances of oxygen and carbon, and in occasional catastrophic bursts fusing the relatively lighter atoms together into the essential heavier elements such as iron, at the same time spewing them back into the interstellar medium from which new, enriched generations of stars and planets could form. The physics of our universe simply requires a vastly old universe—dare I say a venerable universe?—before the path to complex life can begin. And an old universe requires a huge universe—weak as the gravitational forces are, a small universe would collapse gravitationally long before the cosmic cooking had produced a nutritious stew.

Not long ago I sat with Princeton's John Wheeler, the physicist who, among other things, invented the term "black hole." We talked of the way he likened our tiny existence in the vast universe to an immense plant whose central purpose was to produce one small, ephemeral flower. And what an extraordinary efflorescence this is! The human brain is by far the most complex physical object known to us in the entire cosmos. (Only God, "the Old One," the "ground of being" can claim to surpass this complexity.) Of the roughly 35,000 genes coded by the DNA in the human genome, approximately half are expressed in the brain. There are about 100 billion neurons in the brain, nerve cells many of which have long dendritic extensions intricately interconnected with each other. Each neuron connects with about 10,000 other neurons. While the number of estimated stars in all the galaxies in the universe vastly exceeds the grains of sand on all the beaches of the world, the number of synaptic interconnections in a single human brain greatly exceeds the number of stars in our Milky Way: 10^{15} synapses versus 2×10^{11} stars.

For a human at rest, roughly half the body's energy supply fuels the brain. Oxygen (required for slowly "burning" the organic fuel) is carried to the brain by the red blood cells. In these cells the oxygen is loosely bonded to the iron atoms in the middle of the heme complex in the blood's hemoglobin. The oxygen is transferred into the blood through the intricately branched and foliate lung system, where the solubility of oxygen in water and the diameter of the capillaries are finely tuned for an efficient rate of transfer of the oxygen to the heme. Of all the possible metallic complexes, iron has just the right bonding strength to allow the capture and subsequent easy release of the oxygen.

Fortunately for us, our atmosphere contains a reasonable supply of oxygen—about 20 percent by number of atoms. This percentage is high enough to sustain fire, but not so high as to allow cataclysmic combustion. In fact, the acceptable oxygen limits for life more complex than single cells are fairly narrow, and the Earth's atmosphere, it seems, is just right.

As the human brain develops from infancy, a substantial part is devoted to the control of the organs of speech. No other aspect of human powers differs so significantly from the other animals as our ability to communicate by spoken language. It can well be argued that the ability to speak was the first essential step toward becoming human, as Ian Tattersal has done quite eloquently in his book of the same name, *Becoming Human.* The Harvard anthropologist David Pilbeam has remarked that if we could have observed Neanderthal for the 200,000 years beginning that long ago, we could hardly have extrapolated to the complex human civilization that eventually arose on the Earth. The Neanderthals made the same kinds of tools at the beginning until the end of their existence, with no dramatic advance in technology.

Perhaps, as Tattersal has argued, this was owing to their lack of language. Watching the Neanderthals' lack of progress gives us some pause about the inevitability of the evolution of intelligence.

The evidence at hand is hardly conducive to modesty. *Homo sapiens* clearly represents the pinnacle of life on Earth, far outdistancing any rivals, and to say otherwise is to engage in a sort of scholastic fantasy. "What is man that thou are mindful of him?" asks the Psalmist. "For thou hast made him a little lower than the angels and hast crowned him with glory and honor." Yet part of the glory of human creativity and self-consciousness is the ability to ask questions beyond ourselves, about whether the human brain really is the most complex object in the universe or about whether we are alone in the universe—alone in either sense, whether God exists and/or whether extraterrestrial intelligence exists.

Note that the imagined flower on John Wheeler's plant was not only small but ephemeral. The first of the genus *Homo* appeared on Earth about four million years ago, and already half a dozen species of the genus are extinct. *Homo sapiens* has been around only a few hundred thousand years. I asked my neighbor, Philip Morrison, an Institute Professor Emeritus at MIT and an astute observer of the scientific scene, about the prospects for *Homo sapiens*. "I would give it about ten million years," he opined, and when I pressed him about the basis of his estimate, he replied that this was a typical lifetime for a complex species. I agreed that certainly the fossil record shows us that extinction is the name of the game, and there is no reasonable expectation that humankind would be exempt.

While ten million years is a fleeting instant compared to the Sun's projected future lifetime of five billion years, it does seem to me unreasonably long. Given an incredible transition to a sustainable future with a greatly diminished population for our planet, plus a thriving scientific environment needed to harness the vast and generously distributed sunlight, would we not also expect an exponential increase in biological knowledge? Fifty years ago the exact number of chromosomes in human cells was still in doubt, while today we are making a complete map of their genetic patterns. In evolution we have crossed the Lamarkian divide. The human brain now stores more information than the DNA of our chromosomes. What we learn can become inheritable. Correction of crippling genetic defects is close upon us. In another fifty years, barring major nuclear catastrophe, geneticists will surely be able to manipulate the genes to create a stronger, healthier, smarter superman.

There is a rather spooky scenario in Lee Silver's provocative book, *Remaking Eden*.[6] He describes a time three and a half centuries hence when genetic engineering has allowed the wealthier parents to invest in improved genes for their children. Naturally they did not want their GenRich children wasting all this expensive improvement by finding unimproved mates, although inevitably love sometimes crossed those social boundaries between the GenRich and the Naturals. But then an unexpected thing happened: increasingly those mixed marriages proved infertile. I think most of you know the scientific definition of a species, which is the boundary within which individual creatures can interbreed. What Professor Silver has described is the genetic origin of a new species.

In 2,000 years our species will indeed be smarter—if it even exists as the same species! So to me it is unimaginable that *Homo sapiens* will still exist on earth ten million years from now, except perhaps by some remote chance in zoos or special preserves, a throwback much like Przewalski's horse. It is not astronomy that gives me this reading, nor even evolutionary biology and anthropology, but my reflections as a historian and philosopher of science. I believe it is neither pessimistic nor optimistic, but simply realistic. Our universe is going to go on for billions of years without us. Our temporal span is as fleeting as our spatial position is minuscule.

Now as a Christian and as the token theist in your curiously unbalanced exploration of the compatibility between science and religion, I'm much interested in the question, "Is the cosmos all there is?" In making this inquiry I have in mind first of all the question of whether there is a transcendence out of which the universe arose, a creator-designer which accounts for the observed fact that the universe is remarkably well tuned and congenial for the formation of intelligent, self-contemplative consciousness, which in turn may well represent the ultimate purpose of the universe, the bottom line to that most profound of questions, "Why is there something rather than nothing?" Amateur philosopher though I am, I do recognize the anthropomorphic bearings of my question and its answer. "Purpose" is a very human notion that may have no meaning whatsoever to the transcendence that gave birth to the universe. But we are human and I suspect that we will always have to frame our questions and answers in personal, human terms. And I have no trouble imagining that a superintelligence could take on a human mask to interact with our sensibilities, and could communicate to us through prophets of many cultures and many ages.

But when I ask, "Is the Cosmos all there is?" I also have in mind the question of immortality. Does our self-contemplative consciousness have any being beyond this mortal coil? Is our existence just a macabre joke that a meaningless universe has inflicted on us? Is it just sound and fury, signifying nothing? Is it reasonable to hope for a continued existence beyond death, a central doctrine of the Christian faith?

I personally rather like the notion of a personal existence in paradise. I think most of us would enjoy the opportunity to speak again with long-lost friends, to interview our great-grandfathers or Galileo or Aristotle. It would be satisfying, at least fleetingly, to be able to figure out where lost socks went. And I would like the time to learn Greek, and on a longer scale, I would love to look back at the Earth to see how the continents will drift.

But there are problems here. I would like to meet again my mother, who died earlier this year at age ninety-eight, but I would certainly take little pleasure in seeing her at age ninety-eight. We are all continuously changing, generally becoming wiser and more dilapidated. Time is a property of our universe, and a fundamental part of human experience. We live in a world of time, are molded by time, and we cannot envision living in a world without time any more than we can imagine what it would be like to live in a world with time but without growing older.

Last fall I had the opportunity to attend a small workshop on cosmology organized by Martin Rees, specifically on cosmological eschatology, the science of the far future universe, and it was pointed out that none of us have any gut feeling of what "forever" means if indeed we intend an infinity of time, which would allow everything to happen an infinite number of times. As Paul Davies remarked, it is essentially impossible to have a universe that is unending and also interesting. This was, in fact, one of the charges brought against Christianity by Fred Hoyle in his infamous series of radio talks for the BBC half a century ago, the series where he coined the term "Big Bang" as a derisive expression.

I have already suggested that any traditional view of heaven, to the extent it is reasonably well described, is so far from our earthly norms of time and aging that it is tantamount to one of those other universes imagined in the multiverse cosmology. But what modern cosmology is telling us is that it's not automatically absurd to imagine other places with other, unfamiliar physical laws. I suspect our traditional views of heaven are as dated as the tightly nested medieval cosmos. However, it must be as difficult to envision a consistent view of eternity as it is to think of a totally different universe within a multiverse complex, among other reasons because it is hard to grasp the fact that time itself is created with our universe and is no more outside of our universe than is some kind of superspace. Yet, I am told, Plato conceived of a timeless eternity, and I think that is something for philosophers to reexplore. For only in this way, I suspect, will we begin to reconcile our fleeting human existence in this particular universe with the larger cosmological structures that the incredible self-conscious brain of *Homo sapiens* can conceive.

Of course, if you ask me what a timeless eternity is, I cannot answer. Timelessness is as impossible for us to grasp as a beginning of time or an end of time. In a sense it is only wordplay. Nevertheless, word games are not necessarily trivial.

Is the cosmos all there is? The logician will undoubtedly say yes. The philosopher can only say "not necessarily!" He knows he is like the blind man trying to describe an elephant. Is his search for coherence in vain? Or is it a fundamental part of being a reflective, self-conscious creature—maybe even a designed and created creature? Perhaps the quest itself *is* the purpose of the universe and somehow the answer to the question, "Why is there something rather than nothing?" And ultimately, for the Christian, it is a matter of trust.

My canvas today has been vast, digressive, and all too inconclusive. But I hope these fragments may perhaps help you appreciate not only that the endeavor to find a compatibility between science and religion can be intellectually respectable, but that the new cosmology can and should enlarge the vision of theological inquiry in the twenty-first century. In closing, let me turn to a prayer that I have often prayed, the one found near the end of Johannes Kepler's extraordinary cosmological voyage, his *Harmonice mundi:*

> If I have been allured into brashness by the wonderful beauty of thy works, or if I have loved my own glory among men, while advancing in work destined for thy glory, gently and mercifully pardon me: and finally, deign graciously to cause that these demonstrations may lead to thy glory and to the salvation of souls, and nowhere be an obstacle to that. Amen.[7]

Notes

1. The entire debate, along with the presentations of the other speakers at our "Cosmic Questions" symposium, is about to become available from the New York Academy of Sciences, and in a CD-ROM from the AAAS Program on Dialogue between Science, Ethics, and Religion.

2. G. Gamow, *My World Line: An Informal Biography* (New York: Viking Press, 1970), p. 127. Gamow speculated that this parody might account for his not having received an invitation to the 1958 Solvay Congress on cosmology.

3. Fred Hoyle, "The Universe: Past and Present Reflections," pp. 8–12 in *Engineering and Science*, November, 1981, esp. p. 12.

4. Nancy Ellen Abrams and Joel R. Primack, "Scientific Revolutions in Cosmology: Overthrowing vs. Encompassing," essay for a volume edited by Bill Stoeger on philosophical issues in cosmology, in Pachart Press Philosophy in Science series.

5. Christiaan Huygens, *The Celestial Worlds Discover'd* (London, 1698), pp. 154–55.

6. Lee Silver, *Remaking Eden* (New York: Avon Books, 1997), p. 6.

7. End of Book V, chapter 9 of *Harmonice mundi, Johannes Kepler Gesammelte Werke*, 6, 362; my translation is based on the ones by Charles Glenn Wallis in *Great Books of the Western World*, 16, and by Eric J. Aiton, A. M. Duncan, and J. V. Field, *Memoirs of the American Philosophical Society*, 209 (Philadelphia, 1997).

7

BIG-BANG COSMOLOGY AND ATHEISM
Why the Big Bang Is No Help to Theists

QUENTIN SMITH

Since the mid-1960s, scientifically informed theists have been ecstatic because of Big Bang cosmology. Theists believe that the best scientific evidence that God exists is the evidence that the universe began to exist in an explosion about 15 billion years ago, an explosion called the Big Bang. Theists think it obvious that the universe could not have begun to exist uncaused. They argue that the most reasonable hypothesis is that the cause of the universe is God. This theory hinges on the assumption that it is obviously true that whatever begins to exist has a cause.

The most recent statement of this theist theory is in William Lane Craig's 1994 book *Reasonable Faith*.[1] In it Craig states his argument like this:

1. Whatever begins to exist has a cause.
2. The universe began to exist.
3. Therefore, the universe has a cause.[2]

In a very interesting quote from this book he discusses the first premise and mentions me as one of the perverse atheists who deny the obviousness of this assumption:

Quentin Smith's "Big-Bang Cosmology and Atheism: Why the Big Bang Is No Help to Theists" originally appeared in *Free Inquiry* 18, no. 2 (spring 1998).

The first step is so intuitively obvious that I think scarcely anyone could sincerely believe it to be false. I therefore think it somewhat unwise to argue in favor of it, for any proof of the principle is likely to be less obvious than the principle itself. And as Aristotle remarked, one ought not to try to prove the obvious via the less obvious. The old axiom that "out of nothing, nothing comes" remains as obvious today as ever. When I first wrote *The Kalam Cosmological Argument*, I remarked that I found it an attractive feature of this argument that it allows the atheist a way of escape: he can always deny the first premise and assert the universe sprang into existence uncaused out of nothing. I figured that few would take this option, since I believed they would thereby expose themselves as persons interested only in academic refutation of the argument and not in really discovering the truth about the universe. To my surprise, however, atheists seem to be increasingly taking this route. For example, Quentin Smith, commenting that philosophers are too often adversely affected by Heidegger's dread of "the nothing," concludes that "the most reasonable belief is that we came from nothing, by nothing, and for nothing"—a nice ending to a sort of Gettysburg address of atheism, perhaps.[3]

A Baseless Assumption

I'm going to criticize this argument from scientific cosmology, which is the most popular argument that scientifically informed theists and philosophers are now using to argue that God exists.

Let's consider the first premise of the argument, that whatever has a beginning to its existence must have a cause. What reason is there to believe this causal principle is true? It's not self-evident; something is self-evident if and only if everyone who understands it automatically believes it. But many people, including leading theists such as Richard Swinburne, understand this principle very well but think it is false. Many philosophers, scientists, and indeed the majority of graduate and undergraduate students I've had in my classes think this principle is false. This principle is not self-evident, nor can this principle be deduced from any self-evident proposition. Therefore, there's no reason to think it's true. It is either false or it has the status of a statement that we do not know is true or false. At the very least, it is clear that we do not know that it is true.

Now suppose the theist retreats to a weaker version of this principle and says, "Whatever has a beginning to its existence has a cause." Now, this does not say that whatever has a beginning to its existence *must* have a cause; it allows that it is possible that some things begin to exist without a cause. So we don't need to consider it as a self-evident, necessary truth. Rather, according to the theists, we can consider it to be an empirical generalization based on observation.

But there is a decisive problem with this line of thinking. There is absolutely no evidence that it is true. All of the observations we have are of changes in things—of something changing from one state to another. Things move, come to a rest, get larger, get smaller, combine with other things, divide in half, and so on. But we have no observation of things coming into existence. For example, we have no observations of people coming into existence. Here again, you merely have a change of things. An egg cell and a sperm cell change their state by combining. The combination divides, enlarges, and eventually evolves into an adult human being. Therefore, I conclude that we have no evidence at all that the empirical version of Craig's statement, "Whatever begins to exist has a cause," is true. All of the causes we are aware of are changes in preexisting materials. In Craig's and other theists' causal principle, "cause" means something entirely different: it means creating material from nothingness. It is pure speculation that such a strange sort of causation is even possible, let alone even supported in our observations in our daily lives.

An Uncaused Universe

But the more important point is this: not only is there no evidence for the theist's causal assumption, there's evidence against it. The claim that the beginning of our universe has a cause conflicts with current scientific theory. The scientific theory is called the Wave Function of the Universe. It has been developed in the past fifteen years or so by Stephen Hawking, Andre Vilenkin, Alex Linde, and many others. Their theory is that there is a scientific law of nature called the Wave Function of the Universe that implies that it is *highly probable* that a universe with our characteristics will come into existence without a cause. Hawking's theory is based on assigning numbers to all possible universes. All of the numbers cancel out except for a universe with features that our universe possesses, such as containing intelligent organisms. This remaining universe has a very high probability—near 100 percent—of coming into existence uncaused.

Hawking's theory is confirmed by observational evidence. The theory predicts that our universe has evenly distributed matter on a large scale—that is, on the level of superclusters of galaxies. It predicts that the expansion rate of our universe—our universe has been expanding ever since the Big Bang—would be almost exactly between the rate of the universe expanding forever and the rate where it expands and then collapses. It also predicts the very early area of rapid expansion near the beginning of the universe called "inflation." Hawking's theory exactly predicted what the COBE satellite discovered about the irregularities of the background radiation in the universe.[4]

So scientific theory that is confirmed by observational evidence tells us that the universe began without being caused. If you want to be a rational person and accept the results of rational inquiry into nature, then you must accept the fact that God did not cause the universe to exist. The universe exists uncaused, in accordance with the Wave Function law.

Now Stephen Hawking's theory dissolves any worries about how the universe could begin to exist uncaused. He supposes that there is a timeless space, a four-dimensional hypersphere, near the beginning of the universe. It is smaller than the nucleus of an atom. It is smaller than 10^{-33} centimeters in radius. Since it was timeless, it no more needs a cause than the timeless god of theism. This timeless hypersphere is connected to our expanding universe. Our universe begins smaller than an atom and explodes in a Big Bang, and here we are today in a universe that is still expanding.

Is it nonetheless possible that God could have caused this universe? No. For the Wave Function of the Universe implies that there is a 95 percent probability that the universe came into existence uncaused. If God created the universe, he would contradict this scientific law in two ways. First, the scientific law says that the universe would come into existence because of its natural, mathematical properties, not because of any supernatural forces. Second, the scientific law says that the probability is only 95 percent that the universe would come into existence. But if God created the universe, the probability would be 100 percent that it would come into existence because God is all-powerful. If God wills the universe to come into existence, his will is guaranteed to be 100 percent effective.

So contemporary scientific cosmology is not only not supported by any theistic theory, it is actually logically inconsistent with theism.

Notes

1. William Lane Craig, *Reasonable Faith* (Wheaton, Ill.: Crossway, 1994).

2. Ibid., p. 92.

3. Ibid.

4. Confirmation of Hawking's theory is consistent with this theory being a reasonable proposal for the form that an (as yet) undeveloped theory of quantum gravity will take, as Hawking himself emphasizes. See chapter 12, William Lane Craig and Quentin Smith, *Theism, Atheism, and Big Bang Cosmology* (Oxford: Clarendon Press, 1993).

8

HOLY WARS
An Astrophysicist Ponders
the God Question

NEIL DEGRASSE TYSON

At nearly every public lecture that I give on the universe, I try to reserve adequate time at the end for questions. The succession of subjects is predictable. First, the questions relate directly to the lecture. They next migrate to sexy astrophysical subjects such as black holes, quasars, and the Big Bang. If I have enough time left over to answer all questions, and if the talk is in America, the subject eventually reaches God. Typical questions include "Do scientists believe in God?" "Do you believe in God?" and "Do your studies in astrophysics make you more or less religious?"

Publishers have come to learn that there is a lot of money in God, especially when the author is a scientist and when the book title includes a direct juxtaposition of scientific and religious themes. Successful books include Robert Jastrow's *God and the Astronomers*, Leon M. Lederman's *The God Particle*, Frank J. Tipler's *The Physics of Immortality: Modern Cosmology, God, and the Resurrection of the Dead*, and Paul Davies's two works *God and the New Physics* and *The Mind of God*. Each author is either an accomplished physicist

Neil deGrasse Tyson's "Holy Wars: An Astrophysicist Ponders the God Question" appeared in the *Skeptical Inquirer* 25, no. 5 (September/October 2001). It is adapted from an essay that appeared in *Natural History* magazine, October 1999: copyright © Natural History Magazine, Inc., 1999.

or astronomer and, while the books are not strictly religious, they encourage the reader to bring God into conversations about astrophysics. Even Stephen Jay Gould, a Darwinian pitbull and devout agnostic, has joined the title parade with his work *Rocks of Ages: Science and Religion in the Fullness of Life.* The financial success of these published works indicates that you get bonus dollars from the American public if you are a scientist who openly talks about God. After the publication of *The Physics of Immortality,* which suggested whether the law of physics could allow you and your soul to exist long after you are gone from this world, Tipler's book tour included many well-paid lectures to Protestant religious groups. This lucrative subindustry has further blossomed in recent years due to efforts made by the wealthy founder of the Templeton investment fund, Sir John Templeton, to find harmony and consilience between science and religion. In addition to sponsoring workshops and conferences on the subject, Templeton seeks out (among other recipients) widely published religion-friendly scientists to receive an annual award whose cash value exceeds that of the Nobel Prize.

Let there be no doubt that as they are currently practiced, there is no common ground between science and religion. As was thoroughly documented in the nineteenth-century tome *A History of the Warfare of Science with Theology in Christendom,* by the historian and onetime president of Cornell University Andrew D. White, history reveals a long and combative relationship between religion and science, depending on who was in control of society at the time. The claims of science rely on experimental verification, while the claims of religions rely on faith. These approaches are irreconcilable approaches to knowing, which ensures an eternity of debate wherever and whenever the two camps meet. Just as in hostage negotiations, it's probably best to keep both sides talking to each other. The schism did not come about for want of earlier attempts to bring the two sides together. Great scientific minds, from Claudius Ptolemy of the second century to Isaac Newton of the seventeenth, invested their formidable intellects in attempts to deduce the nature of the universe from the statements and philosophies contained in religious writings. Indeed, by the time of his death, Newton had penned more words about God and religion than about the laws of physics, all in a futile attempt to use the biblical chronology to understand and predict events in the natural world. Had any of these efforts succeeded, science and religion today might be largely indistinguishable.

The argument is simple. I have yet to see a successful prediction about the physical world that was inferred or extrapolated from the content of any religious document. Indeed, I can make an even stronger statement. Whenever people have used religious documents to make detailed predictions about the physical world they have been famously wrong. By a prediction I mean a precise statement about the untested behavior of objects or phenomena in the natural world that gets logged *before* the event takes place. When your model predicts something only after it has happened then you have instead made a "postdiction." Postdictions are the backbone of most creation myths and, of course, the "Just So" stories of Rudyard Kipling, where explanations of everyday phenomena explain what is already known. In the business of science, however, a dozen postdictions are barely worth a single successful prediction.

Topping the list of predictions are the perennial claims about when the world will end, none of which have yet proved true. But other claims and predictions have actually stalled or reversed the progress of science. We find a leading example in the trial of Galileo (which gets my vote for the trial of the millennium) where he showed the universe to be fundamentally different from the dominant views of the Catholic Church. In all fairness to the Inquisition, however, an Earth-centered universe made a lot of sense observationally. With a full complement of epicycles to explain the peculiar motions of the planets against the background stars, the time-honored, Earth-centered model had conflicted with no known observations. This remained true long after Copernicus introduced his Sun-centered model of the universe a century earlier. The Earth-centric model was also aligned with the teachings of the Catholic Church and prevailing interpretations of the Bible, wherein Earth is unambiguously created before the Sun and the Moon as described in the first several verses of Genesis. If you were created first, then you must be in the center of all motion. Where else could you be? Furthermore, the Sun and Moon themselves were also presumed to be smooth orbs. Why would a perfect, omniscient deity create anything else?

All this changed, of course, with the invention of the telescope and Galileo's observations of the heavens. The new optical device revealed aspects of the cosmos that strongly conflicted with people's conceptions of an Earth-centered, blemish-free, divine universe: The Moon's surface was bumpy and rocky; the Sun's surface had spots that moved across its surface; Jupiter had moons of its own that orbited Jupiter and not Earth; and Venus went through phases, just like the Moon. For his radical discoveries, which shook Christendom, Galileo was put on trial, found guilty of heresy, and sentenced to house arrest. This was mild punishment when one considers what happened to the monk Giordano Bruno. A few decades earlier Bruno had been found guilty of heresy, and then burned at the stake, for suggesting that Earth may not be the only place in the universe that harbors life.

I do not mean to imply that competent scientists, soundly following the scientific method, have not also been famously wrong. They have. Most scientific claims made on the frontier will ultimately be disproved, due primarily to bad or incomplete data. But this scientific method, which allows for expeditions down intellectual dead ends, also promotes ideas, models, and predictive theories that can be spectacularly correct. No other enterprise in the history of human thought has been as successful at decoding the ways and means of the universe.

Science is occasionally accused of being a closed-minded or stubborn enterprise. Often people make such accusations when they see scientists swiftly discount astrology, the paranormal, Sasquatch sightings, and other areas of human interest that routinely fail double-blind tests or that possess a dearth of reliable evidence. But this same level of skepticism is also being applied to ordinary scientific claims in the professional research journals. The standards are the same. Look what happened when the Utah chemists B. Stanley Pons and Martin Fleischmann claimed in a press conference to create "cold" nuclear fusion on their laboratory table. Scientists acted swiftly and skeptically. Within days of the announcement it was clear that no one could replicate the cold fusion results that Pons and Fleischmann claimed for their experiment. Their work was summarily dismissed. Similar plotlines unfold almost daily (minus the press conferences) for nearly every new scientific claim. The ones that make headlines tend to be the ones that could affect the economy.

With scientists exhibiting such strong levels of skepticism, some people may be surprised to learn that scientists heap their largest rewards and praises upon those who do discover flaws in established paradigms. These same rewards also go to those who create new ways to understand the universe. Nearly all famous scientists, pick your favorite one, have been so praised in their own lifetimes. This path to success in one's professional career is antithetical to almost every other human establishment—especially to religion.

None of this is to say that the world does not contain religious scientists. In a recent survey of religious beliefs among math and science professionals, 65 percent of the mathematicians (the highest rate) declared themselves to be religious, as did 22 percent of the physicists and astronomers (the lowest rate). The national average among all scientists was around 40 percent and has remained largely unchanged over the past century. For reference, 90 percent of the American public claims to be religious (among the highest in Western society), so either nonreligious people are drawn to science or studying science makes you less religious.

But what of those scientists who are religious? One thing is for certain, successful researchers do not get their science from their religious beliefs. On the other hand, the methods of science have little or nothing to contribute to ethics, inspiration, morals, beauty, love, hate, or aesthetics. These are vital elements of civilized life, and are central to the concerns of nearly every religion. What it all means is that for many scientists there is no conflict of interest.

When scientists do talk about God, they typically invoke him at the boundaries of knowledge where we should be most humble and where our sense of wonder is greatest. Examples of this abound. During an era when planetary motions were on the frontier of natural philosophy, Ptolemy couldn't help feeling a religious sense of majesty when he wrote, "When I trace at my pleasure the windings to and fro of the heavenly bodies, I no longer touch the earth with my feet. I stand in the presence of Zeus himself and take my fill of ambrosia." Note that Ptolemy was not weepy about the fact that the element mercury is liquid at room temperature, or that a dropped rock falls straight to the ground. While he could not have fully understood these phenomena either, they were not seen at the time to be on the frontiers of science.

In the thirteenth century, Alfonso the Wise (Alfonso X), the king of Spain who also happened to be an accomplished academician, was frustrated by the complexity of Ptolemy's epicycles. Being less humble than Ptolemy, Alfonso is widely credited with having mused, "Had I been around at the creation, I would have given some useful hints for the better ordering of the universe."

In his 1687 masterpiece *The Mathematical Principles of Natural Philosophy*, Isaac Newton lamented that his new equations of gravity, which describe the force of attraction between pairs of objects, might not maintain a stable system of orbits for multiple planets. Under this instability, planets would either crash into the Sun or get ejected from the solar system altogether. Worried about the long-term fate of Earth and other planets, Newton invoked the hand of God as a possible restoring force to maintain a long-lived solar system. Over a century later, the French mathematician Pierre Simon de Laplace invented a mathematical approach to gravity, published in his four-volume treatise *Celestial Mechanics*, which extended the applicability of Newton's equations to complex systems of planets such as ours. Laplace showed that our solar system was stable and did not require the hand of a deity after all. When queried by Napoleon Bonaparte on the absence of any reference to an "author of the universe" in his book, Laplace replied, "I have no need of that hypothesis."

In full agreement with King Alfonso's frustrations with the universe, Albert Einstein noted in a letter to a colleague, "If God created the world, his primary worry was certainly not to make its understanding easy for us." When Einstein could not figure out how or why a deterministic universe could require the probabilistic formalisms of quantum mechanics, he mused, "It is hard to sneak a look at God's cards. But that he would choose to play dice with the world . . . is something that I cannot believe for a single moment." When an experimental result was shown to Einstein that, if correct, would have disproved his new theory of gravity, Einstein commented, "The Lord is subtle, but malicious he is not." The Danish physicist Niels Bohr, a contemporary of Einstein, heard one too many of Einstein's God-remarks and declared that Einstein should stop telling God what to do!

Today, you hear the occasional astrophysicist (maybe one in a hundred) invoke God when asked where did all our laws of physics come from, or what was around before the Big Bang. As we have come to anticipate, these questions comprise the modern frontier of cosmic discovery and, at the moment, they transcend the answers our available data and theories can supply. Some promising ideas, such as inflationary cosmology and string theory, already exist. These could ultimately give the answers to those questions, thereby pushing back our boundary of awe.

My personal views are entirely pragmatic, and partly resonate with those of Galileo who, during his trial, is credited with saying, "The Bible tells you how to go to heaven, not how the heavens go." Galileo further noted, in a 1615 letter to the Grand Duchess of Tuscany, "In my mind God wrote two books. The first book is the Bible, where humans can find the answers to their questions on values and morals. The second book of God is the book of nature, which allows humans to use observation and experiment to answer our own questions about the universe."

I simply go with what works. And what works is the healthy skepticism embodied in scientific method. Believe me, if the Bible had ever been shown to be a rich source of scientific answers and understanding, we would be mining it daily for cosmic discovery. Yet my vocabulary of scientific inspiration strongly overlaps with that of religious enthusiasts. I, like Ptolemy, am humbled in the presence of our clockwork universe. When I am on the cosmic frontier, and I touch the laws of physics with my pen, or when I look upon the endless sky from an observatory on a mountaintop, I well up with an admiration for its splendor. But I do so knowing and accepting that if I propose a God beyond that horizon, one who graces our valley of collective ignorance, the day will come when our sphere of knowledge will have grown so large that I will have no need of that hypothesis.

II

Intelligent Design
Creationism versus Science

9
CREATIONISM VERSUS EVOLUTION

KENDRICK FRAZIER

Of all the "borderland" areas involving science, the interface between science and religion remains one of the most intriguing and troubling. Scientists, scholars, and laymen continue to ponder the personal and public issues revolving around science and religion. Nearly everyone somehow strives to come to terms both intellectually and emotionally with the array of rich issues involving personal belief on the one hand and commitment to science and reason on the other. Everyone resolves these issues and conflicts in a different way. The spectrum is broad. The issues complex.

At either end of the spectrum, to be sure, beholders have clarity. Evangelical and fundamentalist believers see a black-and-white world. They know the truth. All who do not see it their way are responsible for the world's ills and therefore must be fought with every trick and tactic imaginable. Atheists are equally certain of the correctness of their nonbelief, and everyone else is deluded or at least a bit foolish. Most people are somewhere in between. Most people accommodate a complex system of multilevel, multidimensional, semicompartmentalized beliefs and values.

Kendrick Frazier's "Creationism *vs.* Evolution" originally appeared as "Science and Religion 2001" in the *Skeptical Inquirer* 25, no. 5 (September/October 2001).

That is true of many scientists and scientifically oriented people as well—although those involved in science probably do tend to have fewer adherents to blind belief and more who value and appreciate open-minded inquiry.

Many of the issues are private and personal. In the abstract, what you and I believe (or don't) are each our own business and no one else's. Some of the issues are intellectual. Eminent theologians, great philosophers, Nobel laureate scientists have considered them in depth and shared their insights at length. But others have profound effects on the world—on society, on education, on public policy (and, unfortunately in some cultures where the conflicts have often gone to extremes, on life and limb).

The most troublesome example in the United States (which befuddles those elsewhere) is creationism. Creationists and their sympathizers would expunge from our schools even any mention of evolution—*the* central unifying idea of the biological sciences and one of the most beautiful and most powerfully explanatory concepts in the history of science. They do so in part because they mistakenly fear that evolution somehow undermines human values and dignity. Most of us may see that they are wrong about that, but at least we can see why they are so motivated.

Creationism and its latest spiffed-up manifestation, the "Intelligent Design" (ID) movement, have almost nothing to do with real science and real scientific controversies and everything to do with belief-laden personal and religious politics. But their promoters use scientific language and pretend they are presenting politicians, school board members, and the media valid alternative scientific views. All the while they denigrate every value that science holds dear. These values include unmitigated curiosity, a love of learning, a questioning attitude, an abhorrence of ideology and dogma, a commitment to open-minded inquiry, and an honest acknowledgment that all knowledge is tentative and open to revision (a subtle strength opponents portray and exploit as a serious weakness). Another essential value is a determination to let balanced assessments of facts and evidence guide policy judgments rather than using predetermined ideological views to decide which facts and evidence may be allowed to enter.

I was able to see creationist tactics at work firsthand recently when leading ID proponent Phillip Johnson did a whirlwind speaking tour in New Mexico, where I live and work. Johnson is a UC Berkeley law professor, and as critics predicted before his appearances, he showed that he's very clever at using rhetoric and tactics honed in the legal arena to argue a pretended case against evolution. He distorted, trivialized, and mischaracterized modern evolutionary science to a degree I found shameful. He presented a comic-book-like caricature of evolution that would be laughable if it were not so reprehensible. He bashed an entire broad field of vital science, and he was doing so not as an expert in biology or even in science but as a nonscientist author and ideologue.

But if you think this is clearly an instance where scientifically trained people are able to see through his techniques and realize the intellectual emptiness of the ID argument, you will be surprised. For I heard him at one of the nation's foremost national scientific and engineering laboratories, a huge multiprogram government-funded laboratory that is advancing the frontiers of advanced technology daily, and the overwhelming sentiment of the audience of nearly 400 people there—virtually all scientists and engineers—was on his side. They ate it up. They laughed at his frequent jabs at "materialistic" science, as if their own engineering research was not based on the same science. It was astonishing in a way. In another sense I was not surprised at all.

A glimpse at some of the behind-the-scenes side issues surrounding his appearance shows just how complex and difficult the science and religion issue can be. His invitation to speak did not come from the national lab itself. The lab's upper management was not even aware of his planned appearance until alerted a few weeks before his talk. He was invited by the lab's Christians in the Workforce Networking Group, and most of the attendees were members of the group. The group had been officially sanctioned by the laboratory only as a result of legal action it pressed against the lab for such recognition. The group's official status thus comes under the mandated equal employment opportunity/affirmative action (EEO/AA) part of the lab's administrative operations, not anything to do with science. Furthermore, the group, despite its name, does not represent mainstream Christians at all, but a fundamentalist, evangelical wing. It requires a belief statement to join—ironic, given its EEO/AA home.

Once the lab's management became aware of Johnson's imminent appearance, it found itself in a difficult position. Management didn't like having a person known for antiscience views speak at the lab, but it did not want to be accused of censorship and it did not want to create a controversy that would call attention to Johnson's appearance. It decided to lay low and hope all would pass. Management did require the Christians in the Workforce Group to add a disclaimer to its official lab's Web page. The disclaimer said the talk's location in the lab's main auditorium did not imply any laboratory or government agency "endorsement or approval of any of the concepts or ideas expressed." (This disclaimer was not presented at the talk, however.) In the meantime, a quickly arranged talk by a pro-evolution scientist who some scientists had invited to counter the Johnson talk was canceled by management, on the grounds that that talk didn't have any official sanction—but mainly to avoid overt controversy. Johnson's appearances at other, more public forums in the area got the public attention, and so the lay-low strategy, in a way, worked. But modern biology got roundly bashed at a national laboratory, without refutation.[1]

This example is just a microcosm of how religiously motivated critics of evolution are making inroads in scientific and intellectual arenas. But it wouldn't have happened without a strongly sympathetic potential audience. The example shows that, in the United States at least, scientifically trained people themselves come from a broad spectrum of religious backgrounds, including fundamentalism, and quick generalizations are doomed to failure. If antievolution can be welcomed uncritically in a scientific setting, its acceptance is far easier among other parts of society. Leaders in politics (local to national), education, business, and media are no less diverse and no less vulnerable to distorted arguments against science, if the assertions fit preconceived viewpoints and well-formed mental templates.

The creationist cause continues to be pressed at all levels. In Kansas, where vigilant scientists and educators finally were able to overthrow a creationist takeover of the Kansas State Board of Education, word comes that creationist politicians and supporters are gearing up to retake control. At the national level, a comprehensive U.S. Senate education bill debated for six weeks had attached to it at the last minute a two-sentence amendment drafted by evolution opponents. The innocent-sounding amendment encourages teaching the "controversy" surrounding biological evolution. Its creationist origins are crystal clear: controversies surrounding no other areas of science are singled out. Amidst a flurry of other amendments, the Senate voted 91–8 in favor of the provision on the way to approving the entire education bill by the same margin. Again, a seemingly small inroad, but . . .

Well, the creationist antievolution movement may be among the most pernicious manifestations of conflict between science and religion—or perhaps in this case between good science and bad religion—but related issues, controversies, and concerns are rampant. They always have been, and probably always will be. We're all human, and science and religion, despite their vast differences, are both very human enterprises.

We need to combine a forthright defense of science's highest values with a counseled respect for deeply held personal views. We need to forthrightly deal with all conflicts, without personalizing issues in a way that offends sincere believers who also respect science.

Note

1. After I published these comments in the *Skeptical Inquirer*, the laboratory's director, whom I respect and admire, sent me a short personal note. He said he found some of my characterization of management's views of the controversy "simplistic and uncharitable." Said he: "I have a lot of experience that suggested to me that any attempt to 'correct' someone's religious beliefs on grounds of either logic or science is a fool's errand. It's not worth the time or the effort. We didn't want to expand the 'time wasting' more, by championing a rebuttal." My remarks hadn't been aimed at him and I was sorry that he took some offense, but his comment emphasizes my original point, that even within scientific institutions, dealing with such issues can be exceedingly difficult and complex.

10
SKEPTICISM'S PROSPECTS FOR UNSEATING INTELLIGENT DESIGN

WILLIAM A. DEMBSKI

This conference focuses on skepticism's prospects over the next twenty-five years. I want in this talk specifically to address skepticism's prospects for unseating intelligent design in that time. Though as a proponent of intelligent design I'm no doubt biased, I believe that over the next twenty-five years intelligent design will provide skepticism with its biggest challenge yet. I want in this talk to sketch why I think that.

A few years ago skeptic Michael Shermer wrote a book titled *Why People Believe Weird Things*. Most of the weird things Shermer discusses in that book are definitely on the fringes, like Holocaust denial, alien encounters, and witch crazes—hardly the sort of stuff that's going to make it into the public school science curriculum. Intelligent design by contrast is becoming thoroughly mainstream and threatening to do just that.

Gallup poll after Gallup poll confirms that about 90 percent of the U.S. population believes that some sort of design is behind the world. Ohio is currently the epicenter of the evolution–intelligent design controversy. Recent

William A. Dembski's "Skepticism's Prospects for Unseating Intelligent Design" was originally delivered at CSICOP's Fourth World Skeptics Conference in Burbank, California, on June 21, 2002, at a session entitled "Evolution and Intelligent Design."

polls conducted by the *Cleveland Plain Dealer* found that 59 percent of Ohioans want both evolution and intelligent design taught in their public schools. Another 8 percent want only intelligent design taught. And another 15 percent do not want the teaching of intelligent design mandated, but do want to allow evidence against evolution to be presented in public schools. You do the arithmetic.

Perhaps the most telling finding of this poll is how Ohioans view the consequences for their state of having intelligent design taught in their public schools. According to the *Cleveland Plain Dealer* (June 9[, 2002]), "About three of every four respondents said including intelligent design in the curriculum would have either a positive effect or no effect on the state's reputation or its ability to attract new business." One could hardly imagine the same response if the question were whether to teach astrology, witchcraft, or flat-earth geology. Intelligent design has already become mainstream with the public at large.

Even so, the mainstreaming of intelligent design doesn't cut any ice with skeptics. Skeptics know all about logical and informal fallacies, and the *argumentum ad populum* heads the list. Skepticism purports to keep a gullible public honest. Accordingly, just because intelligent design is acceptable to most Americans doesn't mean that it deserves acceptance (witness America's fascination with horoscopes). All the same, there's reason to think that the usual skeptical assaults are not going to prosper against intelligent design.

One of skepticism's patron saints, H. L. Mencken, remarked, "For every problem, there is a neat, simple solution, and it is always wrong." Yet in writing about Darwin's theory, Stephen Jay Gould remarked, "No great theory ever boasted such a simple structure." Intelligent design claims that Mencken's insight applies to evolutionary biology, overturning not just mechanistic accounts of evolution but skepticism itself.

Skepticism, to be true to its principles, must be willing to turn the light of scrutiny on anything. And yet that is precisely what it cannot afford to do in the controversy over evolution and intelligent design. The problem with skepticism is that it is not a pure skepticism. Rather, it is a selective skepticism that desires a neat and sanitized world which science can in principle fully characterize in terms of unbroken natural laws.

Indeed, why have a skeptical organization with a name like CSICOP? The "COP" in CSICOP is not accidental. CSICOP is in the business of policing claims about the paranormal. The paranormal, by being other than normal, threatens the tidy world bestowed by skepticism's materialistic conception of science.

No other conception of science will do for skepticism. The normal is what is describable by a materialistic science. The paranormal is what's not. Given the skeptic's faith that everything is ultimately normal, any claims about the paranormal must ultimately be bogus. And since intelligent design claims that an intelligence not ultimately reducible to material mechanisms might be responsible for the world and various things we find in the world (not least ourselves), it too is guilty of transgressing the normal and must be relegated to the paranormal.

There is an irony here. The skeptic's world, in which intelligence is not fundamental and the world is not designed, is a rational world because it proceeds by unbroken natural law—cause precedes effect with inviolable regularity. In short, everything proceeds "normally." On the other hand, the design theorist's world, in which intelligence is fundamental and the world is designed, is not a rational world because intelligence can do things that are unexpected. In short, it is a world in which some things proceed "paranormally."

To allow an unevolved intelligence a place in the world is, according to skepticism, to send the world into a tailspin. It is to exchange unbroken natural law for caprice and thereby destroy science. And yet it is only by means of our intelligence that science is possible and that we understand the world. Thus, for the skeptic, the world is intelligible only if it starts off without intelligence and then evolves intelligence. If it starts out with intelligence and evolves intelligence because of a prior intelligence, then the world becomes unintelligible.

The logic here is flawed, but once in its grip, there is no way to escape its momentum. That is why evolution is a nonnegotiable for skepticism. For instance, on two occasions I offered to join the editorial advisory board of Michael Shermer's *Skeptic Magazine* to be its resident skeptic regarding evolution. Though Michael and I are quite friendly, he never took me up on my offer. Indeed, he can't afford to. To do so is to allow that an intelligence outside the world might have influence in the world. That would destroy the world's autonomy and render effectively impossible the global rejection of the paranormal that skepticism requires.

Skepticism therefore faces a curious tension. On the one hand, to maintain credibility it must be willing to shine the light of scrutiny everywhere, and thus in principle even on evolution. On the other hand, to be the scourge with which to destroy superstition and whip a gullible public into line, it must commit itself to a materialistic conception of science and thus cannot afford to question evolution. Intelligent design exploits this tension and thereby turns the tables on skepticism.

Skepticism's love affair with evolution predates Darwin. In fact, it is easily traceable to the atomist and mechanical philosophers of antiquity like Democritus, Epicurus, and Lucretius. Evolution throughout the ages has taught that all aspects of nature, biological complexity included, result from material mechanisms. Within contemporary biology, these include principally the Darwinian mechanism of natural selection and random variation, but also include other mechanisms (symbiosis, gene transfer, genetic drift, the action of regulatory genes in development, self-organizational processes, etc.). These mechanisms are just that: mindless material mechanisms that do what they do irrespective of intelligence. To be sure, mechanisms can be programmed by an intelligence. But any such intelligent programming of evolutionary mechanisms is not properly part of evolutionary theory.

Intelligent design, by contrast, teaches that biological complexity is not exclusively the result of material mechanisms but also requires intelligence, where the intelligence in question is not reducible to such mechanisms. The central issue, therefore, is not the relatedness of all organisms, or what typically is called common descent. Indeed, intelligent design is perfectly compatible with common descent. Rather, the central issue is how biological complexity emerged and whether intelligence played a pivotal role in its emergence.

Suppose, therefore, for the sake of argument that intelligence—one irreducible to material mechanisms—actually did play a decisive role in the emergence of life's complexity and diversity; how could we know it? This question is a special case of a more general question, namely: If an intelligence were involved in the occurrence of some event or the formation of some object, and if we had no direct evidence of such an intelligence's activity, how could we know that an intelligence was involved at all? This last question arises in numerous contexts, including archeology, SETI, and data falsification in science.

I want here to focus on data falsification because it will help point up the legitimacy of the techniques for design detection on which intelligent design depends. On May 23rd of 2001 the *New York Times* reported on the work of "J. Hendrik Schön, 31, a Bell Labs physicist in Murray Hill, N.J., who has produced an extraordinary body of work in the last two and a half years, including seven articles each in *Science* and *Nature*, two of the most prestigious journals." Schön's career is on the line. Why? According to the *New York Times*, Schön published "graphs that were nearly identical even though they appeared in different scientific papers and represented data from different devices. In some graphs, even the tiny squiggles that should arise from purely random fluctuations matched exactly." As a consequence, Bell Labs appointed an independent panel to determine whether Schön "improperly manipulat[ed] data in research papers published in prestigious scientific journals."

The theoretical issues raised in this case of putative data falsification are precisely those that my own work on design detection seeks to address. The match between the two graphs in Schön's articles constitutes an independently given pattern or specification. Moreover, the random fluctuations in the graphs are highly improbable. What's more, the randomness here is well understood. As a consequence, no unknown mechanism is being sought for how the graphs from independent experiments on independent devices could have exhibited the same pattern of random fluctuations. At issue is the question of data manipulation and design, and we resolve it by identifying what I define as "specified improbability" or, as it's also called, "specified complexity."

Regardless whether specified complexity constitutes, as I claim, a sufficient condition for detecting design, it certainly constitutes a necessary condition. Essential to intelligent design is the ability to detect design in cases where the evidence is circumstantial and thus where we lack direct evidence of a designing intelligence. In the case of Schön's graphs, under the relevant chance hypotheses characterizing the random fluctuations in question, the match between graphs had better be highly improbable (if the graphs were merely two-bar histograms with only a few possible gradations in height, then a match between the graphs would be reasonably probable and no one would ever have questioned Schön's integrity). Improbability, however, isn't enough. The random fluctuations of each graph taken individually are indeed highly improbable. But it's the match between the graphs that raises suspicions. That match renders one graph a specification for the other so that in the presence of improbability a design inference is warranted.

By itself detecting design by means of specified complexity does not implicate any particular intelligence. Specified complexity could show that the data in Schön's papers were improperly manipulated. It could not, however, show that Schön was the actual culprit (though as first author on these papers he, like the captain on a proverbial sinking ship, would be in deep trouble). To identify the actual intelligence would require a more thorough causal analysis (an analysis that in the Schön case is being conducted by Bell Labs' independent panel).

I want now to bring this discussion back to the main question of this paper, namely, the prospects for skepticism to unseat intelligent design. To answer this question, let's first consider what intelligent design has going for it:

1. **A method for design detection.** There's much discussion about the validity of specified complexity as a method for design detection, but judging by the response it has elicited over the last four years, this method is not going away. Some scholars (like Wesley Elsberry here) think it merely codifies an argument from ignorance. Others (like Paul Davies) think that it's onto something important. The point is that there are major players who are not intelligent design proponents who disagree. Such disagreement indicates that there are issues of real intellectual merit to be decided and that we're not dealing with a crank theory (at least not one that's obviously so).

2. **Irreducibly complex biochemical systems.** These are systems like the bacterial flagellum. They exhibit specified complexity. Moreover, the biological community does not have a clue how they emerged by material mechanisms. The great promise of Darwinian and other naturalistic accounts of evolution was precisely to show how known material mechanisms operating in known ways could produce all of biological complexity. That promise is now increasingly recognized as unfulfilled (and perhaps unfulfillable). Franklin Harold, not a design proponent, in his most recent book for Oxford University Press, *The Way of the Cell*, states, "There are presently no detailed Darwinian accounts of the evolution of any biochemical or cellular system, only a variety of wishful speculations." Intelligent design contends that our ignorance here comprises not minor gaps in our knowledge of biological systems that promise readily to submit to tried-and-true mechanistic models, but rather indicates vast conceptual lacunae that are bridgeable only by radical ideas like design.

3. **Challenge to the status quo.** Let's face it, in educated circles Darwin-

ism and other mechanistic accounts of evolution are utterly status quo. That has advantages and disadvantages for proponents like yourselves. On the one hand, it means that the full resources of the scientific and educational establishment are behind you, and you can use them to squelch dissent. On the other hand, and especially to the extent that you are heavy-handed in enforcing materialist orthodoxy, it means that you are in danger of alienating the younger generation, which thrives on rebellion against the status quo. Intelligent design appeals to the rebelliousness of youth.

4. **The disconnect between high and mass culture.** It's the educated elite that love evolution and the materialist science it helps to underwrite. On the other hand, the masses are by and large convinced of intelligent design. What's more, the masses ultimately hold the purse strings for the educated elite (in the form of state education, research funding, scholarships, etc.). This disconnect can be exploited. The advantage that biological evolution has had thus far is providing a theoretical framework, however empirically inadequate, to account for the emergence of biological complexity. The disadvantage facing the intelligent-design-supporting masses is that they've had to rely almost exclusively on pretheoretic design intuitions. Intelligent design offers to replace those pretheoretic intuitions with a rigorous design-theoretic framework that underwrites those intuitions.

5. **An emerging research community.** Intelligent design is attracting bright young scholars who are totally committed to developing intelligent design as a research program. We're still thin on the ground, but the signs I see are very promising indeed. It's not enough merely to detect design. Once it's detected, it must be shown how design leads to biological insights that could not have been obtained by taking a purely materialist outlook. I'm beginning to see glimmers of a thriving design-theoretic research program.

What's a skeptic to do against this onslaught, especially when there's a whole political dimension to the debate in which a public tired of being bullied by an intellectual elite find in intelligent design a tool for liberation? Let me suggest the following action points:

1. **Conflate intelligent design with creationism.** I'm not sure how much longer this tactic will work because the public and press are now catching on to the difference, but as long as there's mileage to be obtained, go for it. Emphasize science as a great force for enlightenment and contrast it sharply with fanatical religious fundamentalism. Then stress that intelligent design is essentially a religious and political movement. Generously use the "C-word" to confuse intelligent design with creationism, and then be sure to liken creationism to astrology, belief in a flat earth, and Holocaust denial.

2. **Argue for the superfluity of design.** This action point is also getting increasingly difficult to implement simply on the basis of empirical evidence, but by artificially defining science as an enterprise limited solely to material mechanisms, one conveniently eliminates design from scientific discussion. Thus any gap in our knowledge of how material mechanisms brought about some biological system does not reflect an absence of material mechanisms in nature to produce the system or a requirement for design to account for the system, but only a gap in our knowledge readily filled by carrying on as we have been carrying on.

3. **Play the suboptimality card.** For most people the designer is a benevolent, wise God. This allows for the exploitation of cognitive dissonance by pointing to cases of apparent incompetent or wicked design in nature. I believe intelligent design has good answers to this objection, but the problem of evil is wonderfully adept at clouding intellects. This is one place where skepticism does well exploiting emotional responses.

4. **Achieve a scientific breakthrough.** Provide detailed testable models of how irreducibly complex biochemical systems like the bacterial flagellum could have emerged by material mechanisms. I don't give this much hope, but if you could pull this off, intelligent design would have a lot of backpedaling to do.

5. And finally, **paint a more appealing world picture.** Skepticism is at heart an austere enterprise. It works by negation. It makes a profession of shooting things down. This doesn't set well with a public that delights in novel possibilities. In his *Pensées*, Blaise Pascal wrote, "People almost invariably arrive at their beliefs not on the basis of

proof but on the basis of what they find attractive." Pascal was not talking about people merely believing what they want to believe, as in wish-fulfillment. Rather, he was talking about people being swept away by attractive ideas that capture their heart and imagination. Poll after poll indicates that for most people evolution does not provide a compelling vision of life and the world. Providing such a vision is in my view skepticism's overriding task if it is to unseat intelligent design. Good luck.

11

DESIGN YES, INTELLIGENT NO
A Critique of Intelligent-Design
Theory and Neocreationism

MASSIMO PIGLIUCCI

A new brand of creationism has appeared on the scene in the last few years. The so-called neocreationists largely do not believe in a young Earth or in a too literal interpretation of the Bible. While still mostly propelled by a religious agenda and financed by mainly Christian sources such as the Templeton Foundation and the Discovery Institute, the intellectual challenge posed by neocreationism is sophisticated enough to require detailed consideration (see Edis 2001; Roche 2001).

Among the chief exponents of Intelligent Design (ID) theory, as this new brand of creationism is called, is William A. Dembski, a mathematical philosopher and author of *The Design Inference* (1998a). In that book he attempts to show that there must be an intelligent designer behind natural phenomena such as evolution and the very origin of the universe (see Pigliucci 2000 for a detailed critique). Dembski's (1998b) argument is that modern science ever

Massimo Pigliucci's "Design Yes, Intelligent No: A Critique of Intelligent-Design Theory and Neocreationism" was originally delivered at the conference "Science and Religion: Are They Compatible?" sponsored by the Center for Inquiry, and held in Atlanta, Georgia, in November 2001. It later appeared in the *Skeptical Inquirer* 25, no. 5 (September/October 2001).

since Francis Bacon has illicitly dropped two of Aristotle's famous four types of causes from consideration altogether, thereby unnecessarily restricting its own explanatory power. Science is thus incomplete, and intelligent design theory will rectify this sorry state of affairs, if only close-minded evolutionists would allow Dembski and Co. to do the job.

Aristotle's Four Causes in Science

Aristotle identified *material* causes, what something is made of; *formal* causes, the structure of the thing or phenomenon; *efficient* causes, the immediate activity producing a phenomenon or object; and *final* causes, the purpose of whatever object we are investigating. For example, let's say we want to investigate the "causes" of the Brooklyn Bridge. Its material cause would be encompassed by a description of the physical materials that went into its construction. The formal cause is the fact that it is a bridge across a stretch of water, and not either a random assembly of pieces or another kind of orderly structure (such as a skyscraper). The efficient causes were the blueprints drawn by engineers and the labor of men and machines that actually assembled the physical materials and put them into place. The final cause of the Brooklyn Bridge was the necessity for people to walk and ride between two landmasses while avoiding getting wet.

Dembski maintains that Bacon and his followers did away with both formal and final causes (the so-called teleonomic causes, because they answer the question of *why* something is) in order to free science from philosophical speculation and ground it firmly into empirically verifiable statements. That may be so, but things certainly changed with the work of Charles Darwin (1859). Darwin was addressing a complex scientific question in an unprecedented fashion: he recognized that living organisms are clearly designed in order to survive and reproduce in the world they inhabit; yet, as a scientist, he worked within the framework of naturalistic explanations of such design. Darwin found the answer in his well-known theory of natural selection. Natural selection, combined with the basic process of mutation, makes design possible in nature without recourse to a supernatural explanation because selection is definitely non-random, and therefore has "creative" (albeit non-conscious) power. Creationists usually do not understand this point and think that selection can only eliminate the less fit; but Darwin's powerful insight was that selection is also a cumulative process—analogous to a ratchet—which can build things over time, as long as the intermediate steps are also advantageous.

Darwin made it possible to put all four Aristotelian causes into science. For example, if we were to ask what are the causes of a tiger's teeth within a Darwinian framework, we would answer in the following manner. The material cause is provided by the biological materials that make up the teeth; the formal cause is the genetic and developmental machinery that distinguishes a tiger's teeth from any other kind of biological structure; the efficient cause is natural selection promoting some genetic variants of the tiger's ancestor over their competitors; and the final cause is provided by the fact that having teeth structured in a certain way makes it easier for a tiger to procure its prey and therefore to survive and reproduce—the only "goals" of every living being.

Therefore, design is very much a part of modern science, at least whenever there is a need to explain an apparently designed structure (such as a living organism). All four Aristotelian causes are fully reinstated within the realm of scientific investigation, and science is not maimed by the disregard of some of the causes acting in the world. What then is left of the argument of Dembski and of other proponents of ID? They, like William Paley (1831) well before them, make the mistake of confusing natural design and intelligent design by rejecting the possibility of the former and concluding that any design must by definition be intelligent.

One is left with the lingering feeling that Dembski is being disingenuous about ancient philosophy. It is quite clear, for example, that Aristotle himself never meant his teleonomic causes to imply intelligent design in nature (Cohen 2000). His mentor, Plato (in *Timaeus*), had already concluded that the designer of the universe could not be an omnipotent god, but at most what he called a Demiurge, a lesser god who evidently messes around with the universe with mixed results. Aristotle believed that the scope of god was even more limited, essentially to the role of prime mover of the universe, with no additional direct interaction with his creation (i.e., he was one of the first deists). In *Physics*, where he discusses the four causes, Aristotle treats nature itself as a craftsman, but clearly devoid of forethought and intelligence. A tiger develops into a tiger because it is in its nature to do so, and this nature is due to some *physical* essence given to it by its father (we would call it DNA) which starts the process out. Aristotle makes clear this rejection of god as a final cause (Cohen 2000) when he says that causes are not *external* to the organism (such as a designer would be) but *internal* to it (as modern developmental biology clearly shows). In other words, the final cause of a living being is not a plan, intention, or purpose, but simply intrinsic in the developmental changes of that organism. Which means that Aristotle identified final causes with formal causes as far as living organisms are concerned. He rejected chance and randomness (as do modern biologists) but did not invoke an intelligent designer in its place, contra Dembski. We had to wait until Darwin for a further advance on Aristotle's conception of the final cause of living organisms and for modern molecular biology to achieve an understanding of their formal cause.

Irreducible Complexity

There are two additional arguments proposed by ID theorists to demonstrate intelligent design in the universe: the concept of "irreducible complexity" and the "complexity-specification" criterion. "Irreducible complexity" is a term introduced in this context by molecular biologist Michael Behe in his book *Darwin's Black Box* (1996). The idea is that the difference between a natural phenomenon and an intelligent designer is that a designed object is planned in advance, with forethought. While an intelligent agent is not constrained by a step-by-step evolutionary process, the latter is the only way nature itself can proceed given that it has no planning capacity (this may be referred to as incremental complexity). Irreducible complexity then arises whenever all the parts of a structure have to be present and functional simultaneously for it to work, indicating—according to Behe—that the structure was designed and could not possibly have been gradually built by natural selection.

Behe's example of an irreducibly complex object is a mousetrap. If you take away any of the minimal elements that make the trap work it will loose its function; on the other hand, there is no way to assemble a mousetrap gradually for a natural phenomenon, because it won't work until the last piece is assembled. Forethought, and therefore intelligent design, is necessary. Of course it is. After all, mousetraps as purchased in hardware stores are indeed human products; we *know* that they are intelligently designed. But what of biological structures? Behe claims that, while evolution can explain a lot of the visible diversity among living organisms, it is not enough when we come to the molecular level. The cell and several of its fundamental components and biochemical pathways are, according to him, irreducibly complex.

The problem with this statement is that it is contradicted by the available literature on comparative studies in microbiology and molecular biology, which Behe conveniently ignores (Miller 1996). For example, geneticists are continuously showing that biochemical pathways are partly redundant. Redundancy is a common feature of living organisms where different genes are involved in the same or in partially overlapping functions. While this may seem a waste, mathematical models show that evolution by natural selection has to produce molecular redundancy because when a new function is necessary it cannot be carried out by a gene that is already doing something else, without compromising the original function. On the other hand, if the gene gets duplicated (by mutation), one copy is freed from immediate constraints and can slowly diverge in structure from the original, eventually taking over new functions. This process leads to the formation of gene "families," groups of genes clearly originated from a single ancestral DNA sequence, and that now are diversified and perform a variety of functions (e.g., the globins, which vary from proteins allowing muscle contraction to those involved in the exchange of oxygen and carbon dioxide in the blood). As a result of redundancy, mutations can knock down individual components of biochemical pathways without compromising the overall function—contrary to the expectations of irreducible complexity. (Notice that creationists, never ones to loose a bit, have also tried to claim that redundancy is yet another evidence of intelligent design, because an engineer would produce backup systems to minimize catastrophic failures should the primary components stop functioning. While very clever, this argument once again ignores the biology: the majority of duplicated genes end up as pseudogenes, literally pieces of molecular junk that are eventually lost forever to any biological utility (Max 1986)

To be sure, there are several cases in which biologists do not know enough about the fundamental constituents of the cell to be able to hypothesize or demonstrate their gradual evolution. But this is rather an argument from ignorance, not positive evidence of irreducible complexity. William Paley advanced exactly the same argument to claim that it is impossible to explain the appearance of the eye by natural means. Yet, today biologists know of several examples of intermediate forms of the eye, and there is evidence that this structure evolved several times independently during the history of life on Earth (Gehring and Ikeo 1999). The answer to the classical creationist question, "What good is half an eye?" is "Much better than no eye at all"!

However, Behe does have a point concerning irreducible complexity. It is true that some structures simply cannot be explained by slow and cumulative processes of natural selection. From his mousetrap to Paley's watch to the

Brooklyn Bridge, irreducible complexity is indeed associated with intelligent design. The problem for ID theory is that there is no evidence so far of irreducible complexity in living organisms.

The Complexity-Specification Criterion

William A. Dembski uses an approach similar to Behe to back up creationist claims, in that he also wants to demonstrate that intelligent design is necessary to explain the complexity of nature. His proposal, however, is both more general and more deeply flawed. In his book *The Design Inference* (Dembski 1998a) he claims that there are three essential types of phenomena in nature: "regular," random, and designed (which he assumes to be intelligent). A regular phenomenon would be a simple repetition explainable by the fundamental laws of physics, for example, the rotation of the Earth around the Sun. Random phenomena are exemplified by the tossing of a coin. Design enters any time that two criteria are satisfied: complexity and specification (Dembski 1998b).

There are several problems with this neat scenario. First of all, leaving aside design for a moment, the remaining choices are not limited to regularity and randomness. Chaos and complexity theory have established the existence of self-organizing phenomena (Kauffman 1993; Shanks and Joplin 1999), situations in which order spontaneously appears as an emergent property of complex interactions among the parts of a system. And this class of phenomena, far from being only a figment of mathematical imagination as Behe maintains, is real. For example, certain meteorological phenomena such as tornados are neither regular nor random but are the result of self-organizing processes.

But let us go back to complexity-specification and take a closer look at these two fundamental criteria, allegedly capable of establishing intelligent agency in nature. Following one of Dembski's examples, if SETI (Search for Extra Terrestrial Intelligence) researchers received a very short signal that may be interpreted as encoding the first three prime numbers, they would probably not rush to publish their findings. This is because even though such [a] signal *could* be construed as due to some kind of intelligence, it is so short that its occurrence can just as easily be explained by chance. Given the choice, a sensible scientist would follow Ockham's razor and conclude that the signal does not constitute enough evidence for ET. However, also according to Dembski, if the signal were long enough to encode all the prime numbers between 2 and 101, the SETI people would open the champagne and celebrate all night.

Why? Because such [a] signal would be both too complex to be explained by chance and would be specifiable, meaning that it is not just a random sequence of numbers, it is an intelligible message.

The specification criterion needs to be added because complexity by itself is a necessary but not sufficient condition for design (Roche 2001). To see this, imagine that the SETI staff receives a long but random sequence of signals. That sequence would be very complex, meaning that it would take a lot of information to actually archive or repeat the sequence (you have to know where all the 0s and 1s are), but it would not be specifiable because the sequence would be meaningless.

Dembski is absolutely correct that plenty of human activities, such as SETI, investigations into plagiarism, or encryption, depend on the ability to detect intelligent agency. Where he is wrong is in assuming only one kind of design: for him design equals intelligence and, even though he admitted that such an intelligence may be an advanced extraterrestrial civilization, his preference is for a god, possibly of the Christian variety.

The problem is that natural selection, a natural process, also fulfills the complexity-specification criterion, thereby demonstrating that it is possible to have unintelligent design in nature. Living organisms are indeed complex. They are also specifiable, meaning that they are not random assemblages of organic compounds, but are clearly formed in a way that enhances their chances of surviving and reproducing in a changing and complex environment. What, then, distinguishes organisms from the Brooklyn Bridge? Both meet Dembski's complexity-specification criterion, but only the bridge is irreducibly complex. This has important implications for design.

In response to some of his critics, Dembski (2000) claimed that intelligent design does not mean optimal design. The criticism of suboptimal design has often been advanced by evolutionists who ask why God would do such a sloppy job with creation that even a mere human engineer can easily determine where the flaws are. For example, why is it that human beings have hemorrhoids, varicose veins, backaches, and foot aches? If you assume that we were "intelligently" designed, the answer must be that the designer was rather incompetent—something that would hardly please a creationist. Instead, evolutionary theory has a single answer to all these questions: humans evolved bipedalism (walking with an erect posture) only very recently, and natural selection has not yet fully adapted our body to the new condition (Olshansky et al. 2001). Our closest primate relatives, chimps, gorillas and the like, are better adapted to their way of life, and therefore are less "imperfect" than ourselves!

Dembski is of course correct in saying that intelligent design does not mean optimal design. As much as the Brooklyn Bridge is a marvel of engineering, it is not perfect, meaning that it had to be constructed within the constraints and limitations of the available materials and technology, and it still is subject to natural laws and decay. The bridge's vulnerability to high winds and earthquakes, and its inadequacy to bear a volume of traffic for which it was not built can be seen as similar to the back pain caused by our recent evolutionary history. However, the imperfection of living organisms, already pointed out by Darwin, does do away with the idea that they were created by an omnipotent and omnibenevolent creator, who surely would not be limited by laws of physics that he himself made up from scratch.

The Four Fundamental Types of Design and How to Recognize Them

Given the considerations above, I would like to propose a system that includes both Behe's and Dembski's suggestions, while at the same time showing why they are both wrong in concluding that we have evidence for intelligent design in the universe. Figure 1 summarizes my proposal. Essentially, I think there are four possible kinds of design in nature which, together with Dembski's categories of "regular" and random phenomena, and the addition of chaotic and self-organizing phenomena, truly exhaust all possibilities known to us. Science recognizes regular, random, and self-organizing phenomena, as well as the first two types of design described in figure 1. The other two types of design are possible in principle, but I contend that there is neither empirical evidence nor logical reason to believe that they actually occur.

The first kind of design is *nonintelligent-natural*, and it is exemplified by natural selection within Earth's biosphere (and possibly elsewhere in the universe). The results of this design, such as all living organisms on Earth, are not irreducibly complex, meaning that they can be produced by incremental, continuous (though not necessarily gradual) changes over time. These objects can be clearly attributed to natural processes also because of two other reasons: they are never optimal (in an engineering sense) and they are clearly the result of historical processes. For example, they are full of junk, nonutilized or underutilized parts, and they resemble similar objects occurring simultaneously or

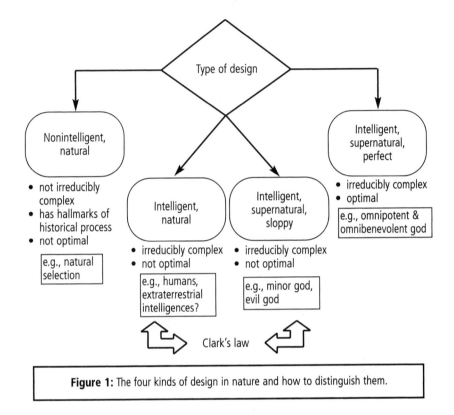

Figure 1: The four kinds of design in nature and how to distinguish them.

previously in time (see, for example, the fossil record). Notice that some scientists and philosophers of science feel uncomfortable in considering this "design" because they equate the term with intelligence. But I do not see any reason to embrace such limitation. If something is shaped over time—by whatever means—such that it fulfills a certain function, then it is designed and the question is simply of how such design happened to materialize. The teeth of a tiger are clearly designed to efficiently cut into the flesh of its prey and therefore to promote survival and reproduction of tigers bearing such teeth.

The second type of design is *intelligent-natural.* These artifacts are usually irreducibly complex, such as a watch designed by a human. They are also not optimal, meaning that they clearly compromise between solutions to different problems (trade-offs) and they are subject to the constraints of physical laws, available materials, expertise of the designer, etc. Humans may not be the only ones to generate these objects, as the artifacts of any extraterrestrial civilization would fall into the same broad category.

The third kind of design, which is difficult, if not impossible, to distinguish from the second, is what I term *intelligent-supernatural-sloppy*. Objects created in this way are essentially indistinguishable from human or ET artifacts, except that they would be the result of what the Greeks called a Demiurge, a minor god with limited powers. Alternatively, they could be due to an evil omnipotent god that just amuses himself with suboptimal products. The reason *intelligent-supernatural-sloppy* design is not distinguishable from some instances (but by all means not all) of *intelligent-natural* design is Arthur C. Clarke's famous third law: from the point of view of a technologically less advanced civilization, the technology of a very advanced civilization is essentially indistinguishable from magic (such as the monolith in his *2001: A Space Odyssey*). I would be very interested if someone could suggest a way around Clarke's law.

Finally, we have *intelligent-supernatural-perfect* design, which is the result of the activity of an omnipotent and omnibenevolent god. These artifacts would be both irreducibly complex and optimal. They would not be constrained by either trade-offs or physical laws (after all, god created the laws themselves). While this is the kind of god many Christian fundamentalists believe in (though some do away with the omnibenevolent part), it's quite clear from the existence of human evil as well as of natural catastrophes and diseases that such [a] god does not exist. Dembski recognizes this difficulty and, as I pointed out above, admits that his intelligent design could even be due to a very advanced extraterrestrial civilization, and not to a supernatural entity at all (Dembski 2000).

Conclusions

In summary, it seems to me that the major arguments of Intelligent Design theorists are neither new nor compelling. (a) It is simply not true that science does not address all Aristotelian causes, whenever design needs to be explained. (b) While irreducible complexity is indeed a valid criterion to distinguish between intelligent and nonintelligent design, these are not the only two possibilities, and living organisms are not irreducibly complex (e.g., see Shanks and Joplin 1999). (c) The complexity-specification criterion is actually met by natural selection, and cannot therefore provide a way to distinguish intelligent from nonintelligent design. (d) If supernatural design exists at all (but where is the evidence or compelling logic?), this is certainly not of the kind that most religionists would likely subscribe to, and it is indistinguishable from the technology of a very advanced civilization.

Therefore, Behe's, Dembski's, and other creationists' (e.g., Johnson 1997) claims that science should be opened to supernatural explanations and that these should be allowed in academic as well as public school curricula is unfounded and based on a misunderstanding of both design in nature and of what the neo-Darwinian theory of evolution (Mayr and Provine 1980) is all about.

Acknowledgments

I would like to thank Melissa Brenneman, Will Provine, and Niall Shanks for insightful comments on earlier versions of this article, as well as Michael Behe, William Dembski, Ken Miller, and Barry Palevitz for indulging in correspondence and discussions with me over these matters.

References

Behe, M. J. 1996. *Darwin's Black Box: The Biochemical Challenge to Evolution.* New York: Free Press.

Cohen, S. M. 2000. "The Four Causes." Accessed on May 16, 2000, Web page, faculty. washington.edu/smcohen.

Darwin, C. [1859] 1910. *The Origin of Species by Means of Natural Selection: Or, the Preservation of Favored Races in the Struggle for Life.* Reprint, New York: A. L. Burt.

Dembski, W. A. 1998a. *The Design Inference.* Cambridge, UK: Cambridge University Press.

———. 1998b. "Reinstating Design within Science." *Rhetoric & Public Affairs* 1: 503–18.

———. 2000. "Intelligent Design Is Not Optimal Design." Accessed on February 3, 2000, Web page, www.meta-list.org.

Edis, T. 2001. "Darwin in Mind: Intelligent Design Meets Artificial Intelligence." *Skeptical Inquirer* 25, no. 2: 35–39.

Gehring, W. J., and K. Ikeo. 1999. "*Pax 6,* Mastering Eye Morphogenesis and Eye Evolution." *Trends in Genetics* 15: 371–77.

Johnson, P. 1997. *Defeating Darwinism by Opening Minds.* Downers Grove, Ill.: InterVarsity Press.

Kauffman, S. A. 1993. *The Origins of Order*. New York: Oxford University Press.

Max, E. E. 1986. "Plagiarized Errors and Molecular Genetics: Another Argument in the Evolution-Creation Controversy." *Creation/Evolution* 9: 34–46.

Mayr, E., and W. B. Provine. 1980. *The Evolutionary Synthesis: Perspectives on the Unification of Biology*. Cambridge, Mass.: Harvard University Press.

Miller, K. R. 1996. "The Biochemical Challenge to Evolution." Accessed on October 30, 1999, Web page, biomed.brown.edu/faculty/M/Miller.

Olshansky, S. J., A. C. Bruce, and R. N. Butler. 2001. "If Humans Were Built to Last." *Scientific American*, March.

Paley, W. 1831. *Natural Theology: Or, Evidences of the Existence and Attributes of the Deity, Collected from the Appearances of Nature*. Boston: Gould, Kendall, and Lincoln.

Pigliucci, M. 2000. "Chance, Necessity, and the New Holy War against Science." Review of *The Design Inference*, by W. A. Dembski. *BioScience* 50, no. 1 (January): 79–81.

Roche, D. 2001. "A Bit Confused: Creationism and Information Theory." *Skeptical Inquirer* 25, no. 2: 40–42.

Shanks, N., and K. H. Joplin. 1999. "Redundant Complexity: A Critical Analysis of Intelligent Design in Biochemistry." *Philosophy of Science* 66: 268–82.

12

THE 'SCIENCE AND RELIGION MOVEMENT'

An Opportunity for Improved Public Understanding of Science?

EUGENIE C. SCOTT

What are we talking about when we talk about "science" and "religion"? Science is a way of knowing that attempts to explain the natural world using natural causes. It is agnostic toward the supernatural—it neither confirms nor rejects it. So science is methodologically materialistic: matter, energy, and their interactions are used to explain nature. Supernatural causes are ruled out for philosophical as well as practical reasons: science requires testing of explanations against the natural world, and testing requires that some variables be held constant. Supernatural forces by definition cannot be held constant, thus supernatural explanation is outside of what science can deal with. Mostly, methodological materialism is embraced by scientists because it works so well; we have found out a great deal about how the universe operates. To say "God did it" does not lead us to greater understanding and tends to discourage further research. Even conservative theologian Alvin Plantinga agrees that resorting to direct supernatural causes to explain the natural world is a "science stopper" (Plantinga 1997).

Eugenie C. Scott's "The 'Science and Religion Movement': An Opportunity for Improved Public Understanding of Science" originally appeared in the *Skeptical Inquirer* 23, no. 4 (July/August 1999).

As an anthropologist, I define religion as a set of rules and beliefs a people have about a nonmaterial universe and its inhabitants. These may include gods, ancestors, powerful spirits, and other supernatural forces. Usually religion includes ideas about an afterlife, but not always. Religion often but not always includes rules about how people should treat one another (ethics and morals). Religion often but not always includes explanations of the natural world. Religious beliefs almost always include a sense of the "spiritual"—awe, wonder, reverence, faith, and other emotions. (Most Americans are Christians, and although "religion" obviously is far broader than just Christianity, my discussion must for reasons of space be limited to this tradition.)

It would appear that science and religion have little in common, yet in the late 1990s there is substantial activity taking place between them. The American Association for the Advancement of Science has an office to promote "Dialogue Between Science and Religion," and in November of 1997 it hosted a major national conference in Chicago titled "The Epic of Evolution." Physical, biological, and social scientists were teamed with theologians to discuss the scientific and theological implications of evolution. Dozens (maybe even scores) of "science and religion" conferences have been held since then, including a large "Science and the Spiritual Quest" conference held on the campus of the University of California sponsored by the Berkeley-based Center for Theology and the Natural Sciences (CTNS). *Science, Newsweek, Time,* and *US News and World Report* have also covered what has come to be called the "science and religion movement."

To some, this is a puzzling development. After all, isn't religion supposed to be in conflict with science, and aren't scientists all secularists? Apparently not. A strong core of scientists are believers. In 1914 the sociologist James H. Leuba surveyed scientists listed in *American Men and Women of Science* and found that 42 percent believed in a personal God—much less than in the general public, but still a substantial number. He predicted that through time, fewer scientists would believe in God—but when Leuba's (albeit problematic) questionnaire was readministered to a group of scientists in 1996, researchers found no appreciable change in the number of "believing" scientists—about 39 percent (Larson and Witham 1997). Many scientists don't see religion and science as inherently incompatible.

In fact, this incompatibility view is found in only one of four ways that (Christian) religion and science historically have interacted.

1. The "warfare" model, as illustrated in Andrew D. White's 1896 classic *A History of the Warfare of Science with Theology in Christendom*, presents religion and science as being incompatible. This perspective is echoed today by Phillip Johnson, Richard Dawkins, Paul Kurtz, and many others. Depending on which side of the issue one is on, one concludes either that religion trumps science, or that science trumps religion.

2. The "separate realms" model understands science and religion to focus on different areas of human concern, with science explaining the natural world, and religion dealing with spiritual matters. Here, science and religion don't conflict, because they have little to say to one another. Stephen Jay Gould is a proponent of this view.

3. The "accommodation" model, in which science and religion are more directly engaged; theological understanding is thought to be deepened through the understanding of science. Some Christians wrestling with the theological implications of Darwinism in the early twentieth century, for example, were willing to reinterpret basic concepts of the Fall, Atonement, and Original Sin in the light of evolutionary theory. These theologians were considering such problems as "If humans evolved from apes, there was no original state of grace and the concept of Original Sin must be reinterpreted" (Bowler 1999, 39). The accommodation seems to be largely a one-way street, with science acting as a source for theological reinterpretation rather than the reverse.

4. In the "engagement" model, science and religion interact as equal partners, stimulating each other to ask different questions than they otherwise might, with the idea that the interaction of both epistemologies will contribute to a fuller understanding of both the natural and nonmaterial realms. This view is reflected in the quotation often attributed to Einstein that "Religion without science is crippled, while science without religion is lame." Scientist and ordained minister Robert J. Russell, of CTNS, said in his introduction to the "Science and the Spiritual Quest" (SSQ) conference:

> For some scientists, the universe as such is the answer. It alone is our source, and science offers us sufficient meaning and purpose. For other scientists, many of whom are gathered here today, science is part of the answer, but a truly adequate account requires language about the God whom Jews, Christians, and Moslems praise as the Creator of the universe and the ultimate source of meaning and purpose in our lives and world. The primary purpose of SSQ is to explore this second option.

. . . Many scholars now see theological doctrines, like scientific theories, not as rigid, closed dogmas but as hypotheses about the world which, while firmly believed to be true, are radically subject to testing by the appropriate data. For at least some of these theologians, the "data" should now include the theories and discoveries of the natural sciences. They also see science as infused with concepts and assumptions whose roots, though often unacknowledged, lie in philosophy and, more indirectly, in Western monotheism, and which invite a critical discussion between theologians, philosophers, and scientists. (Russell 1998, 27)

Most scholars in the "science and religion movement" would identify with the "accommodation" and "engagement" schools, and though some are theologically conservative, very few of them are from biblical literalist Christian traditions. The mainstream science and religion conferences do not include creation science proponents such as Henry Morris and Duane Gish of the Institute for Creation Research. But even neo-creationists like Phillip Johnson and others in the "intelligent design" movement are absent from these conferences. These and other conservative Christians are found in a "parallel universe" such as the "Naturalism, Theism and the Scientific Enterprise" conference at the University of Texas-Austin in February 1997. They promote a very different kind of integration of religion and science called "theistic science," which thus far has been shunned by the mainstream science and religion movement. "Theistic science" is an effort to move science away from methodological materialism and allow in the occasional supernatural explanation— especially for topics such as evolution that have theological implications (Scott 1998). In my opinion, it would be the supreme "science stopper."

There is only one science, but there are many religious views. The science and religion movement is varied because Christian theology is varied. Individuals identifying themselves as Christian hold a variety of concepts of what it means to be a Christian. In my public presentations I often show slides of sign boards from two different Christian churches that vividly illustrate contrasting views of the role (and value) of reason. One, from a fundamentalist church, reads, "Guard Your Mind Or It Will Rule You," a chilling attitude to any lover of reason. The other, from a liberal Protestant church reads, "He Came To Take Away Your Sins, Not Your Brains."

People identifying themselves as Christian also hold a variety of views of God. Some hold to the familiar creator God of special creationism—a Sistine

Chapel–like anthropomorphic figure who creates Adam in his literal likeness. The traditional view of God as omniscient, omnipotent, and omnibenevolent is likely held by most Christians, but theologians and ministers can be found who doubt one or more of those characteristics. There are Christians who see God as a pervasive power, grander than Michelangelo's anthropomorphic figure, but not especially personal (as one minister described it to me, with a twinkle in his eye, "The Force be with you!"). Others who consider themselves religious, if not exactly Christian, have a pantheistic view of "God" as the universe itself, including the laws of the universe. Scientists active in the science and religion movement can be found holding each of these perspectives, and others as well.

My particular interest in science is evolution, and polls clearly show that almost half of Americans deny that evolution occurred. About 40 percent think that evolution occurred, and was guided by God, and an additional 10 percent or so agree that evolution occurred, but deny any role for God. Antievolutionism in the United States and Canada is sustained by the idea that evolution and religion (Christianity) are incompatible; creationists such as Henry Morris demand that people choose between evolution (which is equated with atheism) and Christianity. This is a false dichotomy, of course, as illustrated by the many religious scientists and clergy who accept that evolution occurred. The mainline component of the science and religion movement takes for granted that evolution occurred, and the majority endorse methodological materialism rather than supernatural intervention to explain the natural world. In my observations of the movement, I have seen no support for doctrines such as creation science.

I believe that the science and religion movement will help the public understand the broader range of opinions about evolution among religious individuals, and in this manner improve the climate for the acceptance of evolution as valid science that should be taught in public schools. The theistic science movement, however, needs careful monitoring, as its growth would be detrimental to science as a whole, and to the public understanding of science.

References

Bowler, P. J. 1999. "Going to Extremes in America." Review of *Darwinism Comes to America*, by R. Numbers. *Science* 283: 39.

Larson, Edward J., and Larry Witham. 1997. "Scientists Are Still Keeping the Faith." *Nature* 386: 435–36.

Plantinga, Alvin. 1997. "Methodological Naturalism, Part 2." *Origins and Design* 18, no. 2 (fall): 22–34.

Russell, Robert J. 1998. "Introduction to 'Science and the Spiritual Quest' Conference." *Reports of the National Center for Science Education* 18, no. 2: 26–27.

Scott, Eugenie C. 1998. "'Science and Religion,' 'Christian Scholarship,' and 'Theistic Science': Some Comparisons." *Reports of the National Center for Science Education* 18, no. 2: 30–32.

13

A WORLD DESIGNED BY GOD
Science and Creationism in Contemporary Islam

TANER EDIS

Science and Religion, East and West

In the industrialized West, science and religion interact in a complex manner. While there is potential conflict due to the supernatural perspective of religion and the naturalism of modern science, the *institutions* of Western science and liberal versions of Judaism and Christianity coexist happily. Many good scientists are quite religious as well. This institutional compatibility is strengthened by cultural separation: our conventional wisdom places science and religion in separate spheres.

This picture of limited intellectual conflict and overall social compatibility, however, does not describe science and religion in the Muslim world. There, modern science has not developed as an indigenous heresy, and in the social realm, science is entangled with concerns about Western influence and the perceived necessity of defending Islamic culture. So the relationship

Taner Edis's "A World Designed by God: Science and Creationism in Contemporary Islam" was originally delivered at the conference "Intelligent Design: Are They Compatible?" sponsored by the Center for Inquiry, and held in Atlanta, Georgia, in November 2001.

Figure 1. Muslim apologetics and expositions of the faith often present a harmonious universe: sublime images of galaxies, or a peaceful meadow with butterflies, and couple it with verses from the Quran. This is from a Web site, *Wonders of Creation*, which pair this picture with "Do not the Unbelievers see that the heavens and the earth were joined together (as one unit of creation), before we clove them asunder? We made from water every living thing. Will they not then believe?" (21:30).

between science and religion in the Muslim world is more unsettled, and occasionally strained.

Science in the Quran

Modern science first impressed Muslims through the extremely powerful technology the industrialized West developed and used to dominate Muslim countries. Denying the effectiveness of science was not an option for Muslims; the task was to appropriate science while keeping cautious about other Western influences.

Attempts to fit modern science into an Islamic view of the world, therefore, draw on what are considered to be authentically Islamic intellectual resources. These attempts typically try to preserve a sense of the harmony and design in nature. Muslim apologetic literature, for example, emphasizes classical metaphysical proofs of God less than the Christian tradition. Instead, Muslim writers tend to rely on a commonsense version of the argument from design. They consider it blindingly obvious that the universe exhibits a purposeful harmony; only a fool or a malicious person would overlook the divine hand behind it. Defending the faith means reminding listeners of the obvious existence of God, and then pointing them toward the perfection of the Quran, which must of course with similar obviousness be the uncorrupt Word of God.

This perception of obviousness has roots in the Quran itself, which takes a similar approach to convince its audience of its divine nature. God has given plenty of "signs" in nature to convince the honest doubter:

> It is God who raised the skies without support, as you can see, then assumed His throne, and enthralled the sun and the moon (so that) each runs to a predetermined course. He disposes all affairs, distinctly explaining every sign that you may be certain of the meeting with your Lord. (21:19; Ahmed Ali translation)

Popular apologetics often supplements this attitude with a direct appeal to science. Especially among modernizing sects, claims that certain verses in the Quran anticipate today's scientific knowledge are popular. Muslims take the Quran itself to be the primary miracle Muhammad presented, so establishing that a document originating among desert dwellers fourteen centuries ago contains information only verifiable with modern techniques would be quite impressive. Also drawing on the traditional belief that all useful knowledge is at least rooted in the Quran, such miracle stories have spread far and wide in the Muslim world.

For example, the Quran is supposed to refer to the expanding universe. Some of the verses in question are

> We built the heavens by Our authority; and We are the Lord of power and expanse. We spread the earth a carpet; and what comfort we provide! (51:47, 48)

In some contemporary Muslim eyes, ancient Near-Eastern beliefs about a god spreading the earth, anchoring it with mountains, and establishing a habitation for humans becomes a statement of expanding spacetime. Interestingly, the best-known book about these Quranic miracles is by a French medical doctor (Bucaille 1979). Even a Western scientist, it would seem, has come to acknowledge that the seemingly backward Islamic world had the most important knowledge all along.

Science-in-the-Quran and naïve design arguments can be embarrassing for Muslims used to working in Western or Westernized academic settings (Akhtar 1990). However, another alternative in academic circles is to advocate an "Islamic Science." They take Western science to be marred by materialist presuppositions, and look to classical Islamic philosophy to help construct an approach which is more in tune with revealed spiritual realities.

Though the ideas behind "Islamic Science" are supposed to equally apply to physical science and engineering (Bakar 1999), they have been more influential in the social sciences (e.g., IIIT 1989; Sahin 2001). This is due to a postmodern intellectual climate in these disciplines on one hand, and the political success of Islamists on the other, affecting state universities in the Muslim world.

Curiously, even the disreputable science-in-the-Quran apologetics finds its way into the academic realm. For example, some Saudi medical textbooks detail how the Quran tells us about embryonic development (Moore et al. 1992). So both in the academic and popular arenas of the Muslim world, science is not as culturally separated from religion as it is in the West.

Turkish Creationism

The ambiguous position of science in Islamic culture becomes further clarified when we observe the widespread support for even blatantly pseudoscientific enterprises. The example of Turkish creationism is very illuminating here, since Turkey, being the most Westernized among Muslim countries, is also the stage for some of the most interesting culturally defensive reactions against Western science.

Antievolution sentiment in Turkey is not a new phenomenon. Conservative Muslims have always written against evolution (e.g., Akbulut 1980), and academics with strong links to religious orders can be counted upon to occasionally produce attempts at "Alternative Biology" (e.g., Yilmaz and Uzunoglu 1995). Even those Muslim theologians who declare that Islam is compatible with evolution, or go so far as to claim that Muslim thinkers anticipated evolution long before Darwin was born (Bayrakdar 1987), assume a picture of progressive, guided development, usually in the context of the hierarchically structured realities of classical Islamic philosophy, or the Great Chain of Being. Darwinian evolution, with its naturalistic, accident-driven changes, is rarely referred to except in caricature.

However, with the conservative military regime which took power in

1980, and the subsequent religious-influenced governments, a qualitatively new creationism burst on the scene. In a climate of suppression of secular-leftist politics, and military authorities looking to religion to supply an ideology of national unity, Islamists controlling the Education Ministry took the opportunity to act against evolution. Curiously, they discovered and promoted the Protestant creationism in the United States, even officially translating material from the Institute of Creation Research (ICR) to be used as supplemental reading in state schools (Edis 1994).

Creationism at the state level took up-and-down turns during the politically volatile 1990s, but in the last five years another new development has changed the picture once again. A new face of creationism has appeared, targeting the public through the mass media rather than through formal education. The "Science Research Foundation" (Bilim Arastirma Vakfi [BAV]), an appendage of a religious order led by Adnan Oktar, has publicized creationism through a number of well-produced books and videos under the pseudonym of "Harun Yahya," and a series of "international conferences" in which Turkish creationist academics were joined by American creationists associated with the Institute for Creation Research (Edis 1999; Sayin and Kence 1999).

BAV's creationism has been notable not just because of their media-savvyness, but also because of the considerable resources they command. They are able to make their materials—which are quite attractively put together—available quite cheaply, to the extent of distributing thousands of creationist booklets to passersby on major squares and to students at high school gates. The sources of their finances are quite unclear, as is typical with most Islamist organizations; BAV claims to be supported through donations. Their resources and major media access is the envy of U.S. creationist organizations like the ICR, who state that their connection with BAV is limited to exchange of information and being invited to BAV conferences (John Morris, personal communication). Turkish creationism is certainly more politically *successful* than the U.S. variety.

BAV has lately slowed down slightly in Turkey, because of legal troubles Adnan Oktar encountered due to accusations that his order was using coercive recruiting practices. However, their activity continues, particularly in the area of international outreach. Harun Yahya books, articles, videos, and Web materials (including complete book texts for free) are now available in English, Malay, Russian, Italian, Spanish, Serbo-Croatian (Bosniak), Polish, and

Albanian. Translations into Urdu and Arabic were said to be forthcoming at the time of writing. Harun Yahya books are beginning to become available in Islamic bookstores worldwide, especially as English translations are being printed in London, the center of the Islamic publishing world. Browsers in such bookstores, or visitors to Web sites like www.hyahya.org, will encounter a rich variety of rationally dubious material by Harun Yahya, including Holocaust denial and Masonic conspiracy theories.

It is no accident that large-scale Islamic creationism originated in Turkey, and that BAV material has first been translated into Western languages rather than languages of the Islamic heartland like Arabic and Urdu. The perceived need for creationism is greatest in partially Westernized countries like Turkey, and in the Muslim immigrant communities in the West—evolutionary ideas have to be widely available first before a religious reaction ensues. In any case, Harun Yahya has become popular throughout the Muslim world; he is no longer just a Turkish phenomenon.

Comparing Creationisms

Islamic creationism is rich in ironies for the external observer. It is bad enough that Islamic countries are notoriously invisible in the world's scientific production, but it now looks like Muslims have to borrow even their pseudosciences from the West. Indeed, the lack of originality in those parts of Turkish creationists' writings which attempt to make scientific points is striking. In most cases they lift their arguments from the Western antievolutionist literature, be it young-earth creationism or "Intelligent Design." Nevertheless, Islamic creationism is not an exact copy of the Protestant version, and the differences are as interesting as the similarities.

- In both cases, the most visible form of creationism only gives the illusion of scholarship—misrepresenting evidence, quoting out of context, using dubious sources, etc. However, Islamic creationism enjoys more academic support. Some theologians in major institutions, and even a minority of scientists are sympathetic and publicly supportive of BAV's work. It is as if a number of professors at the Harvard Divinity School were to endorse the ICR.
- Again, in both cases, the creationists' primary concern is apologetic and culturally defensive. Creationism is geared to a popular, already

religious market. Even so, Islamic creationism is part of very successful re-Islamization movement. Creationists have influenced education policies, and institutions such as BAV are very wealthy and politically well connected.

• Both forms of creationism use a stereotyped series of arguments: gaps in the fossil record, the second law of thermodynamics precluding evolution, paleontological fraud, the improbability of complex structure. . . . But BAV does not copy indiscriminately. They omit flood geology and the obsession with Genesis, and draw heavily on the Islamic tradition of perceiving a harmonious, complex universe as a clear sign of divine creation.

This raises the question of why Muslims would borrow from the Protestants, especially since the social and doctrinal differences between their religions can be quite significant. To form a plausible answer, we have to notice that fundamentalists are *not* religious traditionalists. Islamists, though strict about basic doctrines, have a radically different approach to religious authority and political legitimacy than that in traditional orthodox Islam (Brown 2000). As with the U.S. version, the constituency of Turkish creationists is a modernizing, often technically sophisticated population which affords cognitive authority to science (Eve and Harrold 1991). For this audience, U.S. creationism provides a ready-made populist pseudoscience relatively free of Protestant doctrinal idiosyncrasies. Some slight adaptation to an Islamic context is all that is needed.

An important creationist theme which helps understand the motivation of antiscience efforts is their concern to uphold traditional morality in the face of encroaching modernity. Muslim creationists want to defend a moral order revealed in nature as well as scripture. *Fitra*, or "created nature," is central to traditional Islamic conceptions of morality. Right and wrong is closely linked to the well-defined roles people and life-forms have in a specially created universe. Evolution undermines this view of the *nature* of morality: in modern science, biological facts no longer carry immediate moral significance.

Creationism does not exhaust the ways conservative Muslims respond to modern biology. Deploying Islam's own intellectual resources, they occasionally go so far as to resurrect the Aristotelian biology of the Islamic philosophical tradition (Edis and Bix, forthcoming). For example, theologian Süleyman Ates defends traditional sex roles by saying:

It is true that as a whole, the male sex has been created superior to the female. Even the sperm which carries the male sign is different from the female. The male-bearing sperm is more active, . . . the female less. The egg stays stationary, the sperm seeks her out, and endures a long and dangerous struggle in the process. Generally in nature, all male animals are more complete, more superior compared to their females. . . . Man, being more enduring at work, and superior in prudence and willpower, has been given the duty of protecting woman. (Ates 1991; my translation)

Modern science will simply not give conservative Muslims of any stripe what they want. And a liberal Islam similar to the liberal Christianity which self-consciously avoids challenging modern science is barely existent, at least as a movement with a strong constituency.

Conclusion

It is fairly safe to say that the state of science in the Muslim world is dismal. Academic life is too often strangled between lack of resources and political manipulation, and the wider public realm is an arena where all kinds of pseudoscientific claims have a good shot at drowning out scientific responses. In Muslim countries where modernity has made inroads, public distortions of science are, if anything, more widespread. Though any comparison must be unsatisfyingly impressionistic, the situation defenders of science face is probably considerably worse than in the industrialized West. In particular, religiously inspired objections to mainstream science have, on occasion, found significant political and institutional support in the Muslim world.

The continuing conflicts between science and Islam invite some more speculative reflections as well. It would be a mistake to see the relationship of science and religion in Islamic culture as following the West with a time lag of a few centuries. It is not certain that the institutional and cultural conflicts between Islam and science will be resolved in an accommodation like that of liberal Christianity. If anything, the impressive strength of the Islamic revival of the past few decades suggests otherwise. So "Science vs. Religion" debates may come to have an increasingly Islamic flavor in the coming years.

References

Akbulut, S. 1980. *Darwin ve Evrim Teorisi.* Istanbul: Yeni Asya.

Akhtar, S. 1990. *A Faith for All Seasons: Islam and the Challenge of the Modern World.* Chicago: Ivan R. Dee.

Ates, S. 1991. *Gerçek Din Bu,* vol 1. Istanbul: Yeni Ufuklar.

Bakar, O. 1999. *The History and Philosophy of Islamic Science.* Cambridge, UK: Islamic Texts Society.

Bayrakdar, M. 1987. *Islam'da Evrimci Yaradilis Teorisi.* Istanbul: Insan Yayinlari.

Brown, L. C. 2000. *Religion and State: The Muslim Approach to Politics.* New York: Columbia University Press.

Bucaille, M. 1979. *The Bible, the Qur'an and Science.* Indianapolis: American Trust.

Edis, T. 1994. "Islamic Creationism in Turkey." *Creation/Evolution* 34, no. 1.

———. 1999. "Cloning Creationism in Turkey." *Reports of the National Center for Science Education* 19, no. 6: 30.

Edis, T., and A. Bix. Forthcoming. "Premodern Concepts of Gendered Bodies in Current Popular Islam."

Eve, R., and F. Harrold. 1991. *The Creationist Movement in Modern America.* Boston: Twayne.

International Institute of Islamic Thought. 1989. *Islamization of Knowledge.* Herndon: IIIT.

Moore, K. L., et al. 1992. *Human Development as Described in the Quran and Sunnah.* Makkah: Commission on Scientific Signs of the Quran and Sunnah.

Sahin, A. 2001. *Islam ve Sosyoloji açisindan ilim ve din bütünlügü.* Istanbul: Bilge.

Sayin, Ü., and A. Kence. 1999. "Islamic Scientific Creationism." *Reports of the National Center for Science Education,* 19, no. 6: 18.

Yilmaz, I., and S. Uzunoglu. 1995. *Alternatif Biyolojiye Dogru.* Izmir: TÖV.

III

Religion and Science in Conflict

14
SCIENCE AND RELIGION IN HISTORICAL PERSPECTIVE

VERN L. BULLOUGH

The relationship between science and religion, particularly Christianity, is many faceted, and which facets are seen depends on the viewpoint of the authors and the situation in which they find themselves. Views have ranged from regarding the two as mutually interdependent; to independent from each other each with a different sphere of knowledge or *magisterium*; to outright hostility or in terms of Andrew Dickinson White (1896), warfare. The assertion of autonomy of science was central to the stance of the organization of the Royal Society of London (1660) although religious needs powerfully shaped the subsequent articulation of the Newtonian program for natural philosophy (Webster 1975; Jacob 1976). In early Victorian England, William Whewell (1837) and other "broad" church leaders in Anglicanism emphasized that science and theology were distinct from each other but mutually reinforcing (Powell 1834; Cannon 1978). In contrast, Auguste Comte and Karl Marx visioned a science, untrammeled by religious dogma, as providing the key to secular salvation for humanity (Budd 1977). Later, in the United States, John William Draper would advocate a "religion of science," and Andrew Dickinson White would emphasize the conflict between theology and science. Opposing them was the Catholic apologist James Walsh (1908) and the less polemical J. T. Merz (1915). In the last part of the nineteenth century, a synthesis which might have been developing was shattered by Charles Darwin, leading to what can only be called a Victorian crisis of religious faith.

In the twentieth century there were a number of treatises written to assert the importance of metaphysics as a common ground for science and religion (Barnes 1933; Burtt 1925; Eddington 1929; Needham 1925; Whitehead 1925, as well as others). In Britain, one of the leaders of the Humanist movement, Julian Huxley, hoped to create a non-Christian religion that shared some beliefs of the Anglican "Modernists" such as a purposeful universe. He had considerable correspondence with the Anglican Bishop Ernest Barnes on the issue. Ultimately proving to be a stumbling block were Darwin's evolutionary theories. To lessen this George Bernard Shaw proposed his "creative evolution."

Basic to such efforts was the combination of an antimaterialistic interpretation of scientific doctrines with a liberal interpretation of Christian ones. Many of the scientists rejected the alleged materialism and anticlericalism of their nineteenth-century predecessors but held firm to the nineteenth-century belief in progress.

Briefly uniting the two sides were perceived developments in "the new physics," and in non-Darwinian theories of evolution. The proponents of unity held that the apparently indeterministic and even nonmaterial subatomic world revealed by quantum physics could be seen as undermining views cherished by some opponents of religion who had argued that all the phenomena of life and mind were mere epiphenomena of an ultimately deterministic and mechanical material world. To counter the Darwinian theories they put forth Lamarckian ones about the ability to inherit acquired mental and physical characteristics and if this was the case than God could guide the development of humanity.

Such attempts at unity failed in large part because the arguments were grounded in outdated if not outright misleading accounts of the latest scientific developments. Peter Bowler (2001) argues that the mistake of the reconcilers was not so much their reliance on outdated science since even if they had used more current views, these too would soon have been outdated, but depending on science to give them answers to questions about free will, the need for salvation, what humans can know of God and nature, or whether they can save themselves only by appealing to a power beyond themselves.

At the same time these discussions were taking place there were also polemical attempts to demonstrate the importance of "Protestant Christianity" in the formation of modern science (Raven 1942; Hooykaas 1972) as well as sustained historical enquiries into particular conflicts between science and religion (Fleming 1972; Turner 1974; Westfall 1958). Within the sociology of science, there was seen to be a connection between Puritanism and science (Merton 1938; Webster 1975; and Russell 1975).

The point of this brief summation, and it is by no means complete since each recent decade seems to bring an ever-increasing number of books published on the subject, emphasizes that the relationship between science and religion (theology is a better term) is a fertile field. As a historian, I would like to give a kind of overview of the relationships within the Western Christian Church, starting from the beginning. The point I wish to emphasize is that there is not a conflict between science and religion, but there is often one between religion and science. This is because religion tends (gradually often) to incorporate scientific ideas into its theological explanations. The difficulty, as indicated above, is that scientific assumptions do not remain static, but are changing and being modified, occasionally even drastically.

To explain this I would like to start with Aristotle, not Aristotle in the Greek world but Aristotle in the medieval world, since it was the recovery of the Aristotelian corpus in the eleventh, twelfth, and thirteenth centuries which led to the first major conflicts in Western theology over worldviews.

The early Christian Church in the West had a strong intellectual tradition. In fact, one of the reasons it appealed to intellectuals is that it incorporated much of classical learning into its theology. Particularly influential at the time that Western Christian ideas were being formulated were the advocates of Stoicism and Neoplatonism. Concepts from these traditions were incorporated into Christian doctrine by a number of the early church fathers, especially in the writings of St. Augustine (died 430 C.E.), the dominant influence on the Western Church. The Augustinian interpretation of Western theology passed on to the medieval world, but it was only a part of the Greco-Latin corpus of learning. Missing from his theology were the ideas of Aristotle and Greek science in general. Probably the main reason for this was because in the Roman world, much of the Greek scientific tradition remained untranslated into Latin, and the center of science in the Roman world remained primarily in the Eastern parts of the empire at Alexandria and other Greek-speaking centers, and most of the Greek corpus was not easily available. Scientists and scientific investigators such as Galen, the second-century medical writer, continued to write in Greek. What Greek scientific writings that did get translated were for the most part those which could be incorporated into what the Romans called the liberal arts, a term first used by Cicero. Cicero defined them as those studies which were proper for a free man to undertake. There was, however, a certain indecision as to whether there were seven or nine liberal arts. The seven liberal arts consisted of grammar, rhetoric, logic, geometry, arithmetic, astronomy, and music. Those who advocated nine added medicine and architecture.

The number and level of educated individuals who studied the liberal arts (let alone Greek) in the Western Roman Empire rapidly declined during the fifth century when the imperial government in the West fell into the hands of the German chieftains. The chief effect of the German domination was to accentuate the pace of ruralization which had already been underway in the empire. The decline of the city led to a decrease in the number and types of specialists, thus further weakening the whole concept of any kind of specialized education, and lessened the number of schools dedicated to the liberal arts.

Gradually, in place of the Roman schools and institution, the Christian Church established cathedral and monastic schools to keep alive the learning tradition although only that which had managed to be translated into Latin was available. Only a handful of learned men knew Greek and few Greek manuscripts passed directly into the West. In fact, the inheritors of much of the Greek scientific learning were the Muslims, who came to dominate most of the intellectual centers of the eastern parts of the Roman Empire. Islam, tolerant of both Jews and Christians, allowed and encouraged translations of the Greek classics into Arabic, and it was through these translations into Latin that they became part of the Western intellectual tradition. Sometimes there was a roundabout translation with centers in Spain and in southern Italy from Greek to Arabic to Hebrew to Latin. Direct contact was also eventually established with Constantinople, and Greek originals and translations came to the West from there as well. Knowledge and expertise in Greek also grew. The challenge of this new learning to the West led to a major intellectual struggle which was finally settled in the thirteenth century when Albertus Magnus, St. Thomas Aquinas, and others Aristotelianized Christianity and Christianized Aristotle and other Greek scientific writings by incorporating them into their worldview and theology. This had the effect of making the dominant views of Greek science part of Christian doctrine, including the concept that the Sun revolved around the Earth. The acceptance of the new learning was not without struggle, and Aquinas's works for a time were banned at the University of Paris, but eventually they were accepted. Still there were some dissenters. Some of those associated with the newly formed Franciscan order tried to synthesize and reconcile Plato and Aristotle (or the Augustinian version of Plato), emphasizing illumination and a mystical approach.

What is important, in my view, is that by the middle of the thirteenth century, Christian theology had come to be based on what could be regarded as the cutting edge of science at the time. The difficulty that this opposed to Christian theology is that science in its quest for answers is continually growing and changing, and while it is sometimes difficult to change scientific assumptions, those earlier scientific assumptions being challenged had often gained the status of dogma in theological writings. Though there had been serious questions about the heliocentric system by Copernicus and others, the evidence was not clear until the invention of the telescope and the data gathered by Galileo, Kepler, and others. The whole nature of the universe had to be re-conceptualized and new explanations such as Newton's theory of gravity had to be accepted. The result is that the scientific theology of Thomas Aquinas and other Scholastics had been based on what had become outmoded assumptions. The idea of heaven up above and hell and purgatory and other places below had to be rethought, as did a large number of other doctrinal questions.

It took several generations of what might be called Christian apologists to work out ways of reconciliation. In fact it was only in the last years of the twentieth century that the pope formally rehabilitated Galileo, who, incidentally, had always remained a believer in his Church. By this time, however, science had moved far beyond the heliocentric universe to point out that the Sun was only a lesser star in an ever-expanding universe.

How does one incorporate these new discoveries into theological thinking? Should they be incorporated? Should religious belief be based on science? Are they two different magisteria?

The issue, as I indicated, is not with science, but with theology. Every newly developed branch of science poses questions and answers which can be threatening to religion if the two attempt to remain synchronized. As the Darwinian Revolution has gathered steam, only the most rigid dogmatists have opposed the basic premise, but even they have claimed to speak in the name of science, attempting to elevate creationism to the rank of a scientific discipline. Why couldn't they just ignore scientific explanations and indicate that their belief is based on faith? The fact that they refuse to do so and claim that their biblical literalism is a science is what makes for antagonisms between religion and science. Many emphasize that there must be an intelligent design to the universe and it did not happen by chance. They are neither biblical literalists nor hostile critics of science. Some try to reinterpret the Bible by saying that it never states how long a day is, and a lot of things could have taken place before God created light, etc. Some want some kind of stability in belief since it sometimes seems that every new breakthrough in science threatens somebody's religious tenant. The discovery of the sperm and the human ova challenged assumptions which perhaps could be easily revised in theological thinking, but in doing so it leads to new questions such as when did life begin. If the issues were simply ones for theological debate, it would be one thing, but religious assumptions lead to social-political campaigns not only against abortion but also against contraception, since some believe that nothing should intervene to prevent fertilization even though millions of sperm never fertilize an egg and no female can ever have all of her ova fertilized.

Sometimes science, or perhaps only those who speak in the name of science, makes radical changes which even some scientists find difficult to adjust to. Such an issue in today's world is the radical revision of thinking about homosexuality. Perhaps even more radical is the changing role of women in the world at large, and although we still hold to some gender differences, there seems to be a lot of overlap, and traditional views of the inferiority of the female are no longer accepted as scientifically proven. Gradually most bodies of religious believers have made tentative adjustments to scientific explanations, emphasizing that God just worked in natural ways which science can explain. But this raises questions in itself. Where in the universe does God live and work? Where are all those spirits who have gone to heaven or to hell living? Science has no answer to such questions but the problem is that theology feels a need to try to answer such questions without violating the straitjacket which science puts it in.

Still, religion does change. All we need to do is read the Jewish scriptures, which the Christians call the Old Testament, to see what doctrines and practices put forth there are no longer believed or even advocated by even the most dedicated believer. Doctrines prohibiting usury have long since been discarded as we defined a concept of just interest and even state laws forbidding usury have been much modified in recent years. Saturday was originally the Sabbath and Western Christianity long ago abandoned it in favor of the day devoted to the Sun God. If you believe that sparing the rod will spoil the child, you might well end up in jail. We no longer stone people for adultery; most of us don't even practice polygamy. The Christian Church of today, regardless of which particular Christian Church one examines, is not the Christian Church of the past, either immediate or long ago. Even the Old Order Amish who attempt to preserve the customs of the past use horses and buggies which were not known to the early Christians, eat food not found in the Christian Bible, build houses with nails, and in effect, though they deny and abhor modernism, are not living the life of the early Christians.

What I am trying to say is that religious concepts and beliefs have been undergoing a continuous revamping through history even in those religions which claim biblical inerrancy. While representatives of various religious groups individually have represented what might be the best of which human nature is capable of achieving, they also include a significant proportion of humanity engaged in the most evil activities which humans can do. Religion adjusts to life and to reality. Some Christian groups claim it is evil to take human life under any circumstances, but the majority accept war, the death penalty, and even mass exterminations such as occurred with the dropping of the atomic bomb. Often when religions do make changes, they do it kicking and screaming, while all the time making modifications faced with the reality of the world. Some religions can make radical changes apparently without undermining their own positions. An illustration of one of the most radical changes was the Mormon revelation that Blacks who previously had been denied the priesthood and full membership in the Mormon Church could now do so. Changes within Mormonism took place rapidly and with little challenge. Other religions in adjusting seem to disintegrate. The Ambassador Church, headquartered in Pasadena, fell apart when the biblical doctrine it was teaching was changed to meet new standards of biblical criticism. The unwillingness to change, however, guarantees at best little more than survival and increasing isolation, as among the Amish.

Does theology conflict with science? In theory it need not do so; in reality it does if only because it demands a much broader consensus of its constituents for change than the scientific community does. The methods of science, I hold, are the best way of arriving at explanations of phenomenon, but science does not pretend to give all the answers to everything, simply because there is a lot we do not yet know. Religion, however, by its nature, at least in the past, does give answers. The one value that the slower pace of religious change might have for science is the demand for higher standards of proof than science is often able to give, and this demand keeps science on its toes seeking to find new evidence and answers. Unfortunately, when religion strives to stifle this scientific pursuit of answers by prescribing certain beliefs, it only weakens itself.

I have discussed science as a unity, and though it perhaps was once, it is no longer so. In fact, one of the major problems that modern science faces is in keeping track of the many-faceted activities which comprise the scientific endeavor. Science today is highly specialized and in spite of the efforts of such groups as the American Association for the Advancement of Science, and good science teachers in many of the schools and universities, individual scientists know more and more about less and less. Many, in my opinion, even lack an overall view of science. I think it is no accident that those scientists who are the most active in criticizing evolution, usually from a fundamentalist Christian perspective, are engineers or physicians or other professionals, who have a scientific background but confine themselves to working in a very narrow specialty and not thinking about the larger issues. Many, if not most, "scientists" rationalize their own belief system and do not see a conflict with religion because they themselves in their own specialty do not have one and happily belong to churches which do not insist that every word of the Bible is true. The result is that there are "scientists" on all sides of the issue, including vast numbers on the sidelines, and this enables some believers to see it as a battle of my science versus your science. It is not.

Unfortunately, while most scientists I know might be content to let religion go its own way as long as it does not interfere with what they do, those religions which claim a higher power as authority are not always willing to leave either science or the individual scientists alone. Usually it is religious groups doing the pushing, actively trying to deal with "scientific" issues about which they are fearful even though they may know little about them. Thus, though in theory there is not need for conflict, in reality there will be because for many believers, science threatens the very foundation of their belief system. There are a variety of solutions which religious groups have adopted. Some, where the conflict has been deeply embedded in their religious belief system, simply withdraw and seek isolation or partial withdrawal from the world. Examples of such withdrawal are varied from the rigid Orthodox Jews, or devout Muslims, to strict monastic life in the Catholic community or surrounding oneself with a cocoon of protection of fellow believers such as Jehovah's Witnesses. They can not only ignore science, but they also avoid the secularism of the modern world which they believe threatens their values. Another solution, probably adopted by the overwhelming majority, is to rationalize the differences, sort of compartmentalizing their minds and separating science from religion, or picking and choosing what they believe both in their religion and in their science. Still another solution for the dedicated believer is to struggle against the "atheistic scientists" or secular humanists. Comparatively few take this last alternative, but they are the most vocal, and are those now engaged on the front line in the struggle between religion and science. It is the last group which forces science to enter the fray in order to defend itself.

To sum up, it is a war not of religion itself but of a minority of true believers who are interested in imposing their theological views not only on science and scientists but on the world at large. They believe they have the answers and they want the rest of us to accept them. It is a war started by some religious believers not only against science but against secular society itself as well as against other Christians who do not believe as they do. Science does not have all the answers, but religion also leaves many questions unanswered.

References

Barnes, Ernest W. 1933. *Scientific Theory and Religion.* Cambridge: Cambridge University Press.

Bowler, Peter J. 2001. *Reconciling Science and Religion: The Debate in Early Twentieth-Century Britain.* Chicago: University of Chicago Press.

Budd, Susan. 1977. *Varieties of Unbelief: Atheists and Agnostics in English Society, 1850–1960.* London: Heinemann.

Burtt, Edwin A. 1925. *The Metaphysical Foundations of Modern Science.* New York: Harcourt Brace.

Cannon, Susan F. 1978. *Science in Culture: The Early Victorian Period.* New York: Science History Publications.

Draper, John W. 1875. *History of Conflict Between Religion and Science.* New York.

Eddington, Arthur S. 1929. *The Nature of the Physical World.* Cambridge: Cambridge University Press.

Fleming, Donald. 1972. *John William Draper and the Religion of Science.* Reprint. New York: Octagon Books.

Hooykaas, R. 1972. *Religion and the Rise of Modern Science.* Edinburgh: Scottish Academic Press.

Jacob, Margaret C. 1976. *The Newtonians and the English Revolution, 1689–1720.* Ithaca, N.Y.: Cornell University Press.

Merton, Robert K. 1938. "Science, Technology, and Society in Seventeenth-Century England." *Osiris* 4 (Bruges, Belgium: Saint Catherine Press).

Merz, John T. 1896. *A History of European Thought in the Nineteenth Century.* 2 vols., London: Blackwood.

Needham, Joseph, ed. 1925. *Science, Religion, and Reality.* London: Sheldon Press.

Powell, Baden. 1834. *An Historical View of the Progress of the Physical and Mathematical Sciences from the Earliest Ages to the Present Times.* London. New edition. London: Cabinet Cyclopedia, 1837.

Raven, Charles E. *John Ray, Naturalist.* Cambridge: Cambridge University Press.

Russell, Colin, ed. 1973. *Science and Religious Belief: A Selection of Recent Historical Studies.* London: Open University Press.

Turner, Frank M. 1974. *Between Science and Religion: The Reaction to Scientific Naturalism in Late Victorian England.* New Haven: Yale University Press.

Walsh, James T. 1908. *The Popes and Science.* New York: Fordham University Press.

Webster, Charles. 1975. *The Great Instauration: Science, Medicine, and Reform 1616–1660.* London: Duckworth.

Westfall, Richard S. 1958. *Science and Religion in Seventeenth-Century England.* New Haven: Yale University Press.

Whewell, William. 1837. *History of Inductive Sciences: From the Earliest to the Present Time.* 3 vols. London. Reprint. London: Cass, 1967.

White, Andrew Dickinson. 1896. *A History of Warfare of Science with Theology in Christendom.* New York. Reprint. Dover Press, 1960.

Whitehead, Alfred North. 1925. *Science and the Modern World.* New York: Macmillan.

15

THE GALILEO AFFAIR

TIMOTHY MOY

Over the past few decades, historians of science have been reexamining the "Galileo Affair"—Galileo's trial by the Roman Catholic Church in 1633. While scholars have (naturally) been unable to come to a consensus on why Galileo was tried by the Inquisition, almost all historians agree that it was *not* primarily because Galileo believed in Copernican heliocentrism.

The facts of the case are not in dispute. In 1616 Galileo went to Rome to defend his recent writings and public statements promoting heliocentrism after some of his critics had charged that Galileo was promoting a poorly substantiated belief that was contrary to Scripture. By this point, many—perhaps most—Church officials had already concluded that Copernicus's system was the most accurate and useful way to predict astronomical positions (which was particularly important to the Church because of its use in calendar reform), but the question of whether the system was an accurate depiction of reality remained open. First of all, no one had yet come up with a convincing proof that Earth really flew around the Sun at great speed, as Copernicus's proposal required. And second, there were some biblical passages that seemed to suggest

Timothy Moy's "The Galileo Affair" originally appeared as "Science, Religion, and the Galileo Affair" in the *Skeptical Inquirer* 25, no. 5 (September/October 2001).

that Earth was stationary at the center of the universe. This was an unusually touchy subject at the time, since the Church was in the midst of crisis stemming from the Protestant Reformation and was particularly concerned about arguments over who had authority to interpret Scripture.

During his 1616 visit, Galileo received the support of some powerful liberal theologians, particularly Cardinals Roberto Bellarmine and Maffeo Barberini, who argued that, if Copernicus's system was someday proved true, then the Church would have to reinterpret those biblical passages that seemed to contradict it. However, they also supported the compromise that Galileo eventually agreed to: Until such definitive proof was forthcoming, Galileo should discuss heliocentrism only hypothetically, and not promote it as a true description of the heavens.

A Problem of Evidence

Flash forward to 1624. By this time, Galileo had become convinced that he had precisely the proof he was looking for. Even better his old ally, Maffeo Barberini, had recently become Pope Urban VIII. In 1624 Galileo went back to Rome and had six separate audiences with the new pope during which he assured the pontiff that he had worked out a definitive proof of Earth's motion. Urban, intrigued by the prospect of such a demonstration, yet concerned about how the Church would handle the theological consequences, gave Galileo the green light to write about heliocentrism, but still with the understanding that he would not describe it as truth (rather than simply a useful hypothesis) unless he could really prove it.

Convinced that he had the required proof in hand, and feeling that he had the pope's personal blessing to make his case, Galileo published his *Dialogue on the Two Chief World Systems* in 1632. It is a wittily written treatise, crafted as a dialogue between three characters: Simplicio (the geocentric Aristotelian), Salviati (the heliocentric Copernican), and Sagredo (an intelligent and well-informed neutral observer to the debate). In the *Dialogue*, Salviati systematically destroys all of Simplicio's arguments, and concludes with Galileo's new, supposedly conclusive proof that Earth orbits the Sun. Sagredo ultimately concludes that the brilliant Salviati (a transparent stand-in for Galileo himself) is correct, Aristotle is wrong, and everyone retires for wine and snacks.

However, there was one problem: Galileo's new proof made no sense; it was a convoluted argument about how the motion of the tides proves that Earth orbits the Sun, and it simply did not work. When push came to shove (and it did), Galileo simply did not know how to prove that Earth truly moved. Galileo had therefore crossed the line set out sixteen years earlier—he had promoted an idea contrary to Scripture without providing convincing proof of its truthfulness. (In order to protect himself, Galileo had added a preface that claimed that his treatment of heliocentrism was purely hypothetical, but even a casual reading of the *Dialogue* makes clear that this was hogwash; the book was a manifesto for heliocentrism, plain and simple.) Galileo's critics back in Rome instantly seized on the weaknesses of his arguments by charging that Galileo had committed serious offenses: disobeying a papal injunction and promoting teachings contrary to Scripture. (Important note: Galileo was never charged with nor tried for heresy, as is commonly believed. Heresy was a far more serious offense and carried a much stiffer penalty, if you know what I mean.)

In 1633 Galileo was called back to Rome to answer these charges. His trial was a seesaw battle that turned on all manner of technical points in church law, theology, and mathematics, and nearly ended in the equivalent of a hung jury. In the ensuing plea bargain, Galileo admitted that he had gone a bit too far in promoting heliocentrism as truth without sufficient proof and promised not to do it again; all sides then prepared to conclude the face-saving compromise. Then, almost at the last moment (and for reasons that are still quite mysterious), the Inquisition overruled the plea bargain and handed down a verdict and sentence that was unexpectedly harsh: Galileo was found guilty of a "vehement suspicion of heresy" (which was not nearly as bad as heresy itself but still worse than disobedience and teachings contrary to Scripture) and forced to abjure and publicly recant his belief in heliocentrism. Galileo signed a recantation in June of 1633. (I should also point out that Galileo was never imprisoned in a dungeon or tortured during the inquest, as is also sometimes believed. By all accounts, his surroundings were quite enviable.)

After the trial, Galileo returned to his villa outside Florence, where he spent the last decade of his life under comfortable house arrest and injunction not to write anything further on physics. As an indication of how strictly his sentence was carried out, during his remaining years Galileo often stayed at the palaces of nobles and patrons in Tuscany, and openly disobeyed the gag rule by writing his *Discourse on Two New Sciences*, in which he essentially invented kinematics and materials science (though it is true that Galileo's criminal record meant that the book could not be published in Italy; it was published in the Netherlands in 1638). On a purely technical level, the *Discourse* was actually Galileo's greatest contribution to modern science. He died in 1642, the year of Isaac Newton's birth.

Galileo's Punishment

So much for the facts. But why did the Church come down so hard on Galileo? Some scholars argue that Galileo simply had terrible luck, since he happened to be pushing his arguments at the worst possible political moment. In the early seventeenth century, the Catholic Church was desperately trying to fight off an insurrection within Christendom (the Protestant Reformation). Many within the Church hierarchy were not particularly fond of liberalizing Catholic doctrine while it was under assault, and Galileo may have ended up as a collateral casualty of a much larger war.

Other historians argue that an enormous amount of the fault was Galileo's. He was, without a doubt, a voracious social and political climber, and his political maneuverings in the Italian Renaissance court system over his career had garnered him many powerful enemies. With his (erroneous) proof of Copernicanism, Galileo apparently hoped to climb the pyramid to the most prestigious court of all: the Vatican itself (he wanted to become official mathematician/astronomer for the pope). He took a gamble on his proof, lost, and suffered the consequences.

Still other scholars suggest that Galileo's downfall resulted from a personal falling out he had with the pope. There is some evidence to support the conclusion that Urban VIII felt personally betrayed by Galileo's false proof, and was irritated to boot that Galileo had put the pope's words from one of their private conversations into the mouth of Simplicio (the simpleton) at the end of the *Dialogue.*

Personally, I suspect that Galileo got into so much trouble for a variety of reasons. First, he thought heliocentrism was true and became an evangelist for the idea; sadly, there is good reason to believe that Copernican heliocentrism was already succeeding within Church hierarchy and would have become an accepted element of doctrine on its own if Galileo had not forced the issue. Second, he felt that the Church needed to reform its entire intellectual structure in order to modernize and protect itself against Protestantism; in particular, Galileo believed that science had to replace theology as the Church's principal mode of understanding, and that accepting Copernicus was a good first step. Third, he felt that he could have the greatest impact in shaping new doctrine at precisely the moment when the Church was feeling weak and on the defensive. And finally, he felt that he, Galileo Galilei, had the authority and brilliance to transform Catholicism in this way. When you read his writings, you get the distinct impression that Galileo believed that expertise in astronomy and mathematics gave him (and all scientists) a special authority to make theological pronouncements and inform Rome how to run the Church. Frankly, it is no surprise that the Inquisition dropped the hammer on him.

Unfortunately, Galileo's trouble with the Church later became a popular archetype for the historical relationship between science and religion. Nothing could be farther from the truth. For most of the medieval and Renaissance periods, and even stretching into the eighteenth-century Enlightenment, the primary supporter of research and teaching in the sciences was the Roman Catholic Church. In fact, one historian of science, John Heilbron, has recently published a book entitled *The Sun in the Church* that documents how the Church, in the aftermath of the Galileo affair, continued to promote research into evidence for heliocentrism, even to the point of turning entire cathedrals into giant pinhole cameras to measure the apparent diameter of the solar disk at various times of the year. By a mathematical quirk, Copernicus's system would actually produce slightly different variations in the Sun's apparent diameter than the old Ptolemaic-Aristotelian system; the experiments run by the Church in the 1650s and 1660s produced measurements that clearly supported Copernicus.

So, even this classic story of conflict between science and religion is far more complex than most people realize. For me, one of the greatest culprits in the tale is something that still plagues us: a confusion of boundaries between these two ways of understanding the world, and the false belief that expertise in one grants an authority to speak in the other.

Further Reading

Biagioli, Mario. 1993. *Galileo, Courtier: The Practice of Science in the Culture of Absolutism.* Chicago: University of Chicago Press.
De Santillana, Giorgio. 1955. *The Crime of Galileo.* Chicago: University of Chicago Press.
Gingerich, Owen. 1982. "The Galileo Affair." *Scientific American* 247, no. 2: 132–43.

16
UNITING THE WORLD— OR DIVIDING IT
Which Outlook Is Truly Universal, which Parochial in the Extreme?

SIR HERMANN BONDI

Much has been said about the alleged conflict between the products of scientific research and religious statements. I have little enthusiasm for adding to the extensive literature on this topic since we have experienced such flexibility (or should I say slipperiness?) on the part of distinguished defendants of orthodox religion that no such conflict can be defined. If, as authoritative religious spokesmen claim, the Big Bang theory of the universe with an age estimate of over 10 billion years "confirms" the biblical account of a creation a few thousand years ago, debate becomes meaningless. I would much rather concentrate on the contrast in method and outlook between science and religion.

Perhaps the most striking feature of science is its universality. Science is first and foremost a social undertaking. The work of a scientist is futile unless it is understood and absorbed by other scientists, whether to test it, criticize it, demolish it, or build on it. What is so remarkable is that there are no barriers of nationality, race, religion, or ideology in this common endeavor. The yardstick of the empirical test is accepted by all. It is perfectly true that when an experimental test goes against the predictions of a particular theory, some scientists will rejoice and

Sir Hermann Bondi's "Uniting the World—or Dividing It: Which Outlook Is Truly Universal, which Parochial in the Extreme?" originally appeared in *Free Inquiry* 18, no. 2 (spring 1998).

others will grieve. Moreover, the grievers will go through the details of the experiment with a fine-tooth comb. But if any lacunae found are plugged, then all will in due course accept the validity of this empirical disproof of the theory, some enthusiastically, some regretfully. The loss will often not be total, for the defunct theory may well have stimulated a new and fruitful line of experimentation.

Such differences of outlook and expectation are common in science. Indeed, disagreement, discussion, and debate are the lifeblood of science and a source of advancing understanding. But only rarely are these differences of opinion correlated with nationality or race, etc. Far more often are they linked to scientific approach and background: theorists may differ for a time from observers; those who use a particular method of experimentation may favor a different picture from those who use another. But the progress of technology ensures that before long the issue can be settled.

From time to time attempts are made to bring ideology into science, such as Hitler's "Aryan physics" or the Soviets' non-Mendelian agricultural science. They have been abject failures.

The "softest" part of science is the choice of topics to be researched. Fashion plays an important role here: once a costly instrument has been built, it is going to be used, whether the scientific justification that led to the decision to construct it has strengthened during the building period or weakened. When an institute or grouping has been established to research a particular, originally very important, topic, its existence may well outlast the validity of that line of investigation.

The Contrast with Religion

So I am far from saying that science is always pursued with perfection, but its universality is not in doubt. It is in this respect that the contrast with religion is so total. No observer of the world scene can fail to appreciate, first, that many (but by no means all) people have a religious faith and, second, that they belong to numerous different and often mutually contradictory religions. Many of these religions are based on an alleged "revelation."

Revelation is a supposed form of knowledge that, to the believer, is different from, and superior to, all other knowledge ("gospel truth," "the Koran is the word of God," etc.). It is indeed absolute faith in a revelation of supposedly universal validity that marks out the great dogmatic religions. Belief in a god undefined by revelation is far too vague a concept for discussion and so not relevant for this essay. If thought about such a god gives comfort to individuals, I, for one, would not wish to argue them out of it. Indeed, if believers look at their faith as a purely personal matter, then there can be no dispute. But many believers (including the leadership of most institutionalized religions) regard their faith, based on revelation, as "The Truth," applicable to all people everywhere and at all times. These persons view everyone who does not share their particular faith as "in error."

The monstrous arrogance of this outlook is hard to stomach. The wide variety of such faiths and their mutual contradictions must mean that at most one of them can be right and that all the others are wrong. It follows logically that the human mind has a tendency to believe, sincerely and often with fervor, something that is false. To think that oneself and one's fellow believers in one's own faith are uniquely exempt from this general weakness is self-centeredness of stupendous magnitude. To assert for one's own faith possession of the universal truth is to assume an attitude of superiority that in other contexts would be viewed as disgusting. In our world the adherents of every religion form only a minority. Emphasizing a religion claiming universal validity elevates one's own faith to a unique position on the globe and relegates all others to the error bin. Yet for some odd reason a religious statement of alleged superiority is not viewed with the same outraged disgust as would a claim of national or racial superiority.

The Baseless Benefits of Religion

In many circles there is a feeling, to me curious, that religion, any religion, is "a good thing" deserving support, even official, public support. It seems to me that there are a number of arguments, all of debatable value, that underlie this attitude. First, there is the supposition that, since several of the various instructions that faiths give to their followers coincide with our common secular ethic, religious believers lead more moral and law-abiding lives than others. There is little, if any, concrete evidence to support this idea. One of the few possibly relevant items of information concerns the religious affiliations of convicted prisoners. In Britain this gives a checkered picture, with atheists and some faiths underrepresented and Roman Catholics markedly overrepresented. In this country it is plausible to suggest that lip service is less common in Catholicism than in some other religions so that this affiliation must be taken at face value. It is thus reasonable to summarize that this statistic gives no foundation to the assumption that religious faith prevents delinquency. I suspect that elsewhere and for other types of evidence there is equally little to support this thesis.

Another argument for favoring religions is that they act as social cement, as clubs. There is, I suspect, some validity in this supposition. It would be worth investigating whether "loners," so disturbing and sometimes dangerous a feature of the social landscape, are less common among religious people (or those brought up in a religion) than among others. Moreover, if the adherent of some denomination moves to a new location, co-believers will provide a ready-made circle of acquaintances.

But these possible advantages are purchased at a huge cost: this social cement has a tendency to separate the adherents of one faith from all others. In an otherwise very united society this may not matter much, but elsewhere religious separatism is a terrible danger. At its most extreme it is seen in certain cults who do everything possible to separate their adherents from the rest of the world. Most people condemn such cults, but find it difficult, if not impossible, to define a borderline between them and the ordinary religions they regard with favor. How, indeed, can anyone oppose such cults while supporting denominational schooling? After all, the very aim of such schools is to embed the young in a particular faith community, implying thereby that they should inhibit their contacts with outsiders.

A third (and very weak) argument encouraging a favorable attitude to religion is that otherwise the smaller denominations are liable to suffer from paranoia that strengthens separatism and hostility to outsiders. Public recognition and perhaps financial support, it is sometimes argued, may make such groups less defensive and withdrawn. The mere mention of the awkward separatist cults demolishes this argument.

Public hostility to religions has been shown to be ineffective and to encourage fanaticism. But public attitudes that firmly reject state involvement in religion and in no way favor religion over irreligion can and do work. Religious separatism is an enemy of civilized society and should be firmly condemned. This attitude is often described, particularly in India, as secularism.

The true contrast between science and religion is that science unites the world and makes it possible for people of widely differing backgrounds to work together and to cooperate. Religion, on the other hand, by its very claim to know "The Truth" through "revelation," is inherently divisive and a creator of separatism and hostility.

17
WHY GETTING IT RIGHT MATTERS
How Science Prevails

DANIEL C. DENNETT

Here is a story you probably haven't heard, about how a team of American researchers inadvertently introduced a virus into a Third World country they were studying. They were experts in their field, and they had the best intentions; they thought they were helping the people they were studying, but in fact they had never really seriously considered whether what they were doing might have ill effects.

The virus they introduced had some dire effects indeed: it raised infant mortality rates, led to a general decline in the health and well-being of women and children, and, perhaps worst of all, indirectly undermined the only effective political force for democracy in the country, strengthening the hand of the traditional despot who ruled the nation. When the researchers were confronted

Daniel C. Dennett's "Why Getting It Right Matters: How Science Prevails" was an invited address to the World Congress of Philosophy, Boston, August 13, 1998. Portions are derived from "Faith in the Truth," Amnesty Lecture, Oxford, February 17, 1997, published as "Postmodernism and Truth" in *The Values of Science: The Amnesty Lectures, Oxford 1997*, edited by Wes Williams (New York: Westview Press/Perseus Books, 1999). Copyright © 1999 by Westview Press. Reprinted with permission of Westview Press, a member of Perseus Books, LLC. This article also appeared in *Free Inquiry* 20, no. 1 (winter 1999/2000).

with the devastation they had wrought, their response was frustrating, to say the least: they still thought that what they were doing was, all things considered, in the interests of the people, and declared that the standards by which this so-called devastation was being measured were simply not appropriate. Their critics, they contended, were trying to impose "Western" standards in a cultural environment that had no use for them. In this strange defense they were warmly supported by the country's leaders—not surprisingly—and little was heard—also not surprisingly—from those who might have been said, by Western standards, to have suffered as a result of their activities.

These researchers were not biologists intent on introducing new strains of rice, nor were they agribusiness chemists testing new pesticides, or doctors trying out vaccines that couldn't legally be tested in the United States. They were postmodernist science critics and other multiculturalists who were arguing, in the course of their professional researches on the culture and traditional "science" of this country, that Western science was just one among many equally valid narratives, not to be "privileged" in its competition with native traditions that other researchers—biologists, chemists, doctors, and others— were eager to supplant. The virus they introduced was not a macromolecule but a meme (a replicating idea): the idea that science was a "colonial" imposition, not a worthy substitute for the practices and beliefs that had carried the Third World country to its current condition. And the reason you have not heard of this particular incident is that I made it up to dramatize the issue and to try to unsettle what seems to be current orthodoxy among the literati about such matters. But it is inspired by real incidents—that is to say, true reports. Events of just this sort have occurred in India and elsewhere and reported, movingly, by a number of writers.[1]

My little fable is also inspired by a wonderful remark of E. O. Wilson, in *Atlantic Monthly* a few months ago: "Scientists, being held responsible for what they say, have not found postmodernism useful." Actually, of course, we are all held responsible for what we say. The laws of libel and slander, for instance, exempt none of us, but most of us—including scientists in many or even most fields—do not typically make assertions that, independently of libel and slander considerations, might bring harm to others, even indirectly. A handy measure of this fact is the evident ridiculousness we discover in the idea of malpractice insurance for literary critics, philosophers, mathematicians, historians, cosmologists, etc. What on earth could a mathematician or literary critic do, in the course of executing her professional duties, that might need the security blanket of malpractice insurance? She might inadvertently trip a student in the corridor, or drop a book on somebody's head, but, aside from such outré side effects, our activities are paradigmatically innocuous. One would think. But in those fields where the stakes are higher—and more direct—there is a long-standing tradition of being especially cautious, and of taking particular responsibility for ensuring that no harm results (as explicitly honored in the Hippocratic Oath). Engineers, knowing that thousands of people's safety may depend on the bridge they design, engage in focused exercises with specified constraints designed to determine that, according to all current knowledge, their designs are safe and sound. Even economists—often derided for the risks they take with other people's livelihoods—when they find themselves in positions to endorse specific economic measures considered by government bodies or by their private clients, are known to attempt to put a salutary strain on their underlying assumptions, just to be safe. They are used to asking themselves, and to being expected to ask themselves: "What if I'm wrong?"

We others seldom ask ourselves this question, since we have spent our student and professional lives working on topics that are, according both to tradition and common sense, incapable of affecting any lives in ways worth worrying about. If my topic is whether or not Vlastos had the best interpretation of Plato's *Parmenides* or how the wool trade affected imagery in Tudor poetry, or what the best version of string theory says about time, or how to recast proofs in topology in some new formalism, if I am wrong, dead wrong, in what I say, the only damage I am likely to do is to my own scholarly reputation. But when we aspire to have a greater impact on the "real" (as opposed to "academic") world—and many philosophers do aspire to this today—we need to adopt the attitudes and habits of these more applied disciplines. We need to hold ourselves responsible for what we say, recognizing that our words, if believed, can have profound effects for good or ill.

Truth versus Passing Fancies

When I was a young, untenured professor of philosophy, I received a visit from a colleague from the comparative literature department, an eminent and fashionable literary theorist, who wanted some help from me. I was flattered to be asked, and did my best to oblige, but the drift of his questions about various philosophical topics was strangely perplexing to me. Finally he managed to make clear to me what he had come for. He wanted "an epistemology," he said. An epistemology. Every self-respecting literary theorist had to sport an epistemology that season, it seems, and without one he felt naked, so he had come to me for an epistemology to wear—it was the very next fashion, he was sure, and he wanted the dernier cri in epistemologies. It didn't matter to him that it be sound, or defensible, or (as one might as well say) true; it just had to be new and different and stylish. Accessorize, my good fellow, or be overlooked at the party.

At that moment I perceived a gulf between us that I had only dimly seen before. It struck me at first as simply the gulf between being serious and being frivolous. But that initial surge of self-righteousness on my part was, in fact, a naive reaction. My sense of outrage, my sense that my time had been wasted by this man's bizarre project, was in its own way as unsophisticated as the reaction of the first-time theater-goer who leaps on the stage to protect the heroine from the villain. "Don't you understand?" we ask incredulously. "It's make-believe. It's art. It isn't supposed to be taken literally!" Put in that context, perhaps this man's quest was not so disreputable after all. I would not have been offended, would I, if a colleague in the drama department had come by and asked if he could borrow a few yards of my books to put on the shelves of the set for his production of Tom Stoppard's play, *Jumpers*. What if anything would be wrong in outfitting this fellow with a snazzy set of outrageous epistemological doctrines with which he could titillate or confound his colleagues?

What would be wrong would be that, since this man didn't acknowledge the gulf, didn't even recognize that it existed, my acquiescence in his shopping spree would have contributed to the debasement of a precious commodity, the erosion of a valuable distinction. Many people, including both onlookers and participants, don't see this gulf, or actively deny its existence, and therein lies the problem. The sad fact is that, in some intellectual circles, inhabited by some of our more advanced thinkers in the arts and humanities, this attitude passes as a sophisticated appreciation of the futility of proof and the relativity of all knowledge claims. In fact this opinion, far from being sophisticated, is the height of sheltered naiveté, made possible only by flat-footed ignorance of the proven methods of scientific truth seeking and their power. Like many another naif, these thinkers, reflecting on the manifest inability of their methods of truth seeking to achieve stable and valuable results, innocently generalize from their own cases and conclude that nobody else knows how to discover the truth either.

Among those who contribute to this problem, I am sorry to say, is my good friend Dick Rorty. Richard Rorty and I have been constructively disagreeing with each other for over a quarter of a century now. Rorty has opened up the horizons of contemporary philosophy, shrewdly showing us philosophers many things about how our own projects have grown out of the philosophical projects of the distant and recent past, while boldly describing and prescribing future paths for us to take. But there is one point over which he and I do not agree at all—not yet—and that concerns his attempt over the years to show that philosophers' debates about Truth and Reality really do erase the gulf, really do license a slide into some form of relativism. In the end, Rorty tells us, it is all just "conversations," and there are only political or historical or aesthetic grounds for taking one role or another in an ongoing conversation.

Rorty has often tried to enlist me in his campaign, declaring that he could find in my own work one explosive insight or another that would help him with his project of destroying the illusory edifice of objectivity. One of his favorite passages is the one with which I ended my book *Consciousness Explained* (1991):

> It's just a war of metaphors, you say—but metaphors are not "just" metaphors; metaphors are the tools of thought. No one can think about consciousness without them, so it is important to equip yourself with the best set of tools available. Look what we have built with our tools. Could you have imagined it without them? (p. 455)

"I wish," Rorty says, "he had taken one step further, and had added that such tools are all that inquiry can ever provide, because inquiry is never 'pure' in the sense of [Bernard] Williams's 'project of pure inquiry.' It is always a matter of getting us something we want."[2] But I would never take that step, for, although metaphors are indeed irreplaceable tools of thought, they are not the only such tools. Microscopes and mathematics and MRI scanners are among the others. Yes, any inquiry is a matter of getting us something we want: the truth about something that matters to us, if all goes as it should.

When philosophers argue about truth, they are arguing about how not to inflate the truth about truth into the Truth about Truth, some absolutistic doctrine that makes indefensible demands on our systems of thought. It is in this regard similar to debates about, say, the reality of time, or the reality of the past. There are some deep, sophisticated, worthy philosophical investigations into whether, properly speaking, the past is real. Opinion is divided, but you entirely misunderstand the point of these disagreements if you suppose that they undercut claims such as the following:

- Life first emerged on this planet more than three thousand million years ago.
- The Holocaust happened during World War II.
- Jack Ruby shot and killed Lee Harvey Oswald at 11:21 A.M., Dallas time, November 22, 1963.

These are truths about events that really happened. Their denials are falsehoods. No sane philosopher has ever thought otherwise, though in the heat of battle they have sometimes made claims that could be so interpreted.

Richard Rorty deserves his large and enthralled readership in the arts and humanities, and in the "humanistic" social sciences, but when his readers enthusiastically interpret him as encouraging their postmodernist skepticism about truth, they trundle down paths he himself has refrained from traveling. When I press him on these points, he concedes that there is indeed a useful concept of truth that survives intact after all the corrosive philosophical objections have been duly entered. This serviceable, modest concept of truth, Rorty acknowledges, has its uses: when we want to compare two maps of the countryside for reliability, for instance, or when the issue is whether the accused did or did not commit the crime as charged.

Even Richard Rorty, then, acknowledges the gap, and the importance of the gap, between appearance and reality, between those theatrical exercises that may entertain us without pretense of truth telling, and those that aim for, and often hit, the truth. He calls it a "vegetarian" concept of truth. Very well, then, let's all be vegetarians about the truth. Scientists never wanted to go the whole hog anyway.

Basic Instinct

So now, let's ask about the sources or foundations of this mild, uncontroversial, vegetarian concept of truth.

Right now, as I speak, billions of organisms on this planet are engaged in a game of hide and seek. But it is not just a game for them. It is a matter of life and death. Getting it right, not making mistakes, has been of paramount importance to every living thing on this planet for more than three billion years, and so these organisms have evolved thousands of different ways of finding out about the world they live in, discriminating friends from foes, meals from mates, and ignoring the rest for the most part. It matters to them that they not be misinformed about these matters—indeed nothing matters more—but they don't, as a rule, appreciate this. They are the beneficiaries of equipment exquisitely designed to get what matters right, but, when their equipment malfunctions and gets matters wrong, they have no resources, as a rule, for noticing this, let alone deploring it. They soldier on, unwittingly. The recognition of the difference between appearance and reality is a human discovery. A few other species—some primates, some cetaceans, maybe even some birds—show signs of appreciating the phenomenon of "false belief"—getting it wrong. They exhibit sensitivity to the errors of others, and perhaps even some sensitivity to their own errors as errors, but they lack the capacity for the reflection required to dwell on this possibility, and so they cannot use this sensitivity in the deliberate design of repairs or improvements of their own seeking gear or hiding gear. That sort of bridging of the gap between appearance and reality is a wrinkle that we human beings alone have mastered.

We are the species that discovered doubt. Is there enough food laid by for winter? Have I miscalculated? Is my mate cheating on me? Should we have moved south? Is it safe to enter this cave? Other creatures are often visibly agitated by their own uncertainties about just such questions, but because they cannot actually ask themselves these questions, they cannot articulate their predicaments for themselves or take steps to improve their grip on the truth. They are stuck in a world of appearances, making the best they can of how things seem.

We alone can be racked with doubt, and we alone have been provoked by that epistemic itch to seek a remedy: better truth-seeking methods. Wanting to keep better track of our food supplies, our territories, our families, our enemies, we discovered the benefits of talking it over with others, asking questions, passing on lore. We invented culture. Then we invented measuring, and arithmetic, and maps, and writing. The point of asking questions is to find true answers; the point of measuring is to measure accurately; the point of making maps is to find your way to your destination. The Land of the Liars could exist only in philosophers' puzzles; there are no traditions of False Calendar Systems for misrecording the passage of time. In short, the goal of truth goes without saying, in every human culture.

We human beings use our communicative skills not just for truth telling, but also for promise making, threatening, bargaining, storytelling, entertaining, mystifying, inducing hypnotic trances, and just plain kidding around, but prince of these activities is truth telling, and for this activity we have invented ever better tools. Alongside our tools for agriculture, building, warfare, and transportation, we have created a technology of truth: science. Try to draw a straight line, or a circle, "freehand." Unless you have considerable artistic talent, the result will not be impressive. With a straight edge and a compass, on the other hand, you can practically eliminate the sources of human variability and get a nice clean, objective result, the same every time.

Is the line really straight? How straight is it? In response to these questions, we develop ever finer tests, and then tests of the accuracy of those tests, and so forth, bootstrapping our way to ever greater accuracy and objectivity. Scientists are just as vulnerable to wishful thinking, just as likely to be tempted by base motives, just as venal and gullible and forgetful as the rest of humankind. Scientists don't consider themselves to be saints; they don't even pretend to be priests (who according to tradition are supposed to do a better job than the rest of us at fighting off human temptation and frailty). Scientists take themselves to be just as weak and fallible as anybody else, but recognizing those very sources of error in themselves and in the groups to which they belong, they have devised elaborate systems to tie their own hands, forcibly preventing their frailties and prejudices from infecting their results.

It is not just the implements, the physical tools of the trade, that are designed to be resistant to human error. The organization of methods is also under severe selection pressure for improved reliability and objectivity. The classic example is the double-blind experiment, in which, for instance, neither the human subjects nor the experimenters themselves are permitted to know which subjects get the test drug and which the placebo, so that nobody's subliminal hankerings and hunches can influence the perception of the results. The statistical design of both individual experiments and suites of experiments is then embedded in the larger practice of routine attempts at replication by independent investigators, which is further embedded in a tradition—flawed, but recognized—of publication of both positive and negative results.

What inspires faith in arithmetic is the fact that hundreds of scribblers, working independently on the same problem, will all arrive at the same answer (except for those negligible few whose errors can be found and identified to the mutual satisfaction of all). This unrivaled objectivity is also found in geometry and the other branches of mathematics, which since antiquity have been the very model of certain knowledge set against the world of flux and controversy. In Plato's early dialogue, the *Meno*, Socrates and the slave boy work out together a special case of the Pythagorean theorem. Plato's example expresses the frank recognition of a standard of truth to be aspired to by all truth-seekers, a standard that has not only never been seriously challenged, but that has been tacitly accepted—indeed heavily relied upon, even in matters of life and death—by the most vigorous opponents of science. (Or do you know a church that keeps track of its flock, and their donations, without benefit of arithmetic?)

Yes, but science almost never looks as uncontroversial, as cut and dried, as arithmetic. Indeed rival scientific factions often engage in propaganda battles as ferocious as anything to be found in politics, or even in religious conflict. The fury with which the defenders of scientific orthodoxy often defend their doctrines against the heretics is probably unmatched in other arenas of human rhetorical combat. These competitions for allegiance—and, of course, funding—are designed to capture attention, and being well-designed, they typically succeed. This has the side effect that the warfare on the cutting edge of any science draws attention away from the huge uncontested background, the dull metal heft of the ax that gives the cutting edge its power. What goes without saying, during these heated disagreements, is an organized, encyclopedic collection of agreed-upon, humdrum scientific fact.

Robert Proctor usefully draws our attention to a distinction between neutrality and objectivity.[3] Geologists, he notes, know a lot more about oil-bearing shales than about other rocks—for the obvious economic and political reasons—but they do know objectively about oil bearing shales. And much of what they learn about oil-bearing shales can be generalized to other, less favored rocks. We want science to be objective; we should not want science to be neutral. Biologists know a lot more about the fruitfly, *Drosophila*, than they do about other insects—not because you can get rich off fruit flies, but because you can get knowledge out of fruit flies easier than you can get it out of most other species. Biologists also know a lot more about mosquitoes than about other insects, and here it is because mosquitoes are more harmful to people than other species that might be much easier to study. Many are the reasons for concentrating attention in science, and they all conspire to making the paths of investigation far from neutral; they do not, in general, make those paths any less objective. Sometimes, to be sure, one bias or another leads to a violation of the canons of scientific method. Studying the pattern of a disease in men, for instance, while neglecting to gather the data on the same disease in women, is not just not neutral; it is bad science, as indefensible in scientific terms as it is in political terms.

It is true that past scientific orthodoxies have themselves inspired policies that hindsight reveals to be seriously flawed. One can sympathize, for instance, with Ashis Nandy, editor of the passionately antiscientific anthology, *Science, Hegemony and Violence: A Requiem for Modernity* (Delhi: Oxford University Press, 1988). Having lived through Atoms for Peace, and the Green Revolution, to name two of the most ballyhooed scientific juggernauts that have seriously disrupted Third World societies, he sees how "the adaptation in India of decades-old western technologies are advertised and purchased as great leaps forward in science, even when such adaptations turn entire disciplines or areas of knowledge into mere intellectual machines for the adaptation, replication and testing of shop-worn western models which have often been given up in the west itself as too dangerous or as ecologically non-viable" (p. 8). But we should recognize this as a political misuse of science, not as a fundamental flaw in science itself.

How Science Prevails

The methods of science aren't foolproof, but they are indefinitely perfectible. Just as important: there is a tradition of criticism that enforces improvement whenever and wherever flaws are discovered. The methods of science, like everything else under the sun, are themselves objects of scientific scrutiny, as method becomes methodology, the analysis of methods. Methodology in turn falls under the gaze of epistemology, the investigation of investigation itself— nothing is off limits to scientific questioning. The irony is that these fruits of scientific reflection, showing us the ineliminable smudges of imperfection, are sometimes used by those who are suspicious of science as their grounds for denying it a privileged status in the truth-seeking department—as if the institutions and practices they see competing with it were no worse off in these regards. But where are the examples of religious orthodoxy being simply abandoned in the face of irresistible evidence? Again and again in science, yesterday's heresies have become today's new orthodoxies. No religion exhibits that pattern in its history.

Notes

1. Meera Nanda, "The Epistemic Charity of the Social Constructivist Critics of Science and Why the Third World Should Refuse the Offer," in *A House Built on Sand: Exposing Postmodernist Myths about Science*, ed. N. Koertge (New York: Oxford University Press, 1998), pp. 286–311. Reza Afshari, "An Essay on Islamic Cultural Relativism in the Discourse of Human Rights," in *Human Rights Quarterly* 16 (1994): 235–76. Susan Okin, "Is Multiculturalism Bad for Women?" *Boston Review*, October/ November, 1997, pp. 25–28. Pervez Hoodbhoy, *Islam and Science: Religious Orthodoxy and the Battle for Rationality* (London and New Jersey: Zed Books Ltd., 1991).

2. "Holism, Intrinsicality, Transcendence," in *Dennett and His Critics*, ed. Bo Dahlbom (Oxford: Blackwells, 1993).

3. *Value-Free Science?* (Cambridge, Mass.: Harvard University Press, 1991).

18
THE DANGEROUS QUEST FOR COOPERATION BETWEEN SCIENCE AND RELIGION

JACOB PANDIAN

Recently, misleading articles have appeared in newspapers and news magazines claiming that religion and science are cooperating to explore the nature of reality. Gregg Easterbrook (1999) noted that "Signs of renewed interest in science and religion are numerous. The topic has recently been a top-selling cover for both *Newsweek* and *U.S. News and World Report.* Universities such as Princeton and Cambridge, which in the 1960s didn't even offer courses in the relationship between science and religion, have established chairs for its study."

Easterbrook points to the central role of the John Templeton Foundation in encouraging the cooperation between science and religion. The foundation publishes *Progress in Theology* magazine but more importantly awards millions of dollars to people who reflect their philosophy of cooperation. The 2001 Templeton prize, $1 million, was announced March 9. It went to the Rev. Arthur Peacocke, a British biochemist and Anglican priest who has written widely about God and science. The Templeton Award recipient for 2000 was Freeman J. Dyson, an emeritus professor of physics at the Institute of Advanced Study in Princeton. As reported by Larry Stammer (2000), Dyson was "baffled"

Jacob Pandian's "The Dangerous Quest for Cooperation Between Science and Religion" originally appeared in the *Skeptical Inquirer* 25, no. 5 (September/October 2001).

at receiving the award because the Templeton prize is awarded for "Progress in Religion" and not for progress in science. Dyson claimed that he was "not a theologian" and "not a saint." In his reflections on science and religion, Dyson noted that "The universe has a mind of its own. We know mind plays a big role in our own lives. It's likely, in fact, that mind has a big role in the way the whole universe functions. If you like, you call it God. It all makes sense."

Before that, $1.2 million was awarded to Ian G. Barbour, a retired professor from Carleton College. At Carleton he was professor of physics, professor of religion, and Bean Professor of Science, Technology and Society. His book *Religion and Science* (1997) is described by its publisher (Harper San Francisco) as "a definitive contemporary discussion of the many issues surrounding our understanding of God and religious truth and experience in our scientific age." Earlier recipients of the Templeton Award include the Protestant Christian evangelist Billy Graham, the Catholic Christian nun Mother Teresa, the campus crusader William Bright, and the Russian novelist Alexander Solzhenitsyn. Ian Barbour, according to Gregg Easterbrook, "promptly announced he would give $1 million of his award to the Berkeley, California, Center for Theology and the Natural Sciences, an affiliate of Berkeley's Graduate Theological Union, and an organization whose own 1981 founding and rising importance are indicators of the science-and-religion trend."

Ralph Estling, in an essay called "Templeton and AAAS" in the *Skeptical Inquirer* (2000), pointed out that the American Association for the Advancement of Science has "a problem": This association, which "has been promoting a study known as the 'Program of Dialogue on Science, Ethics and Religion,'" received for four years cash contributions of over one million dollars from the Templeton Foundation. As many board members of AAAS are also associated with the Templeton Foundation, Estling is right in raising questions about "conflict of interest," and he advises the AAAS "to get the hell out from under the John Templeton Foundation."

I suggest that the problem is a much larger one than the Templeton Foundation's attempt to influence the scope of science through monetary awards to scientific organizations and scientists. The more serious problem stems from our profound misunderstanding of why and how the concept of religion was developed by the church fathers of the early medieval period out of the Roman/Latin concept of *religio*. It is this misunderstanding that opens the door to organizations such as the Templeton Foundation, and to arguments that science and religion should cooperate in understanding the nature of the universe.

Religio, Religion, and Supernaturalism

Supernaturalism (i.e., beliefs and practices associated with supernatural beings and supernatural power) is a cultural universal. Religion, however, is not a cultural universal; it is a subset of supernaturalism that developed during the medieval period of the Christian tradition to represent Christian supernaturalism as scientific truth. During this period, the Roman/Latin concept of *religio* changed its meaning and significance from ritual activities to doctrinal statements about the nature of the world and humankind.

An excellent discussion of why and how the Roman/Latin concept of *religio* was transformed by the church fathers into religion (attributing different characteristics to *religio*) is offered in Wilfred Cantwell Smith's very important book on the subject of religion, *The Meaning and End of Religion* (1991). The concept of religion was developed in the Christian tradition to represent Christian truths as opposed to the untruths of "pagan" traditions of the Greeks and Romans and the satanic or demonic distortions that, from the Christian theory of religion, prevailed in non-Christian traditions.

The concept of religion that had become the theoretical framework for explaining Greco-Roman and non-Western traditions as false was also opposed to and contrasted with the supernaturalism of the non-Christians in general. Christian supernaturalism was conceptualized within the framework of religion as the scientific truth about the world and humankind. Christianity established an epistemological link between science and supernaturalism by conceptualizing religion as the framework to explain natural phenomena and to explain the nature of the relationship between God and humankind. In such a view, the religious framework of Christianity was aligned with scientific naturalism, and non-Christian supernaturalism was aligned with superstition.

For over fifteen hundred years we have been using the term *religion* without fully realizing its origin and development. Scholars have used the term to identify and discuss the supernaturalism of both non-Christian and Christian traditions. But while ancient civilizations such as the Greeks, Romans, Chinese, and Hindus had elaborate beliefs and activities that we associate with supernaturalism, they did not have "religion," i.e., a formulation that combines scientific knowledge and supernaturalism. Thus labels such as Greek religion,

Roman religion, Chinese religion, Hindu religion, and so on are erroneous. It would be more appropriate to discard the use of the term *religion* and instead attempt to define and discuss Christian supernaturalism, just as we describe and discuss other supernaturalisms.

Is there conflict or cooperation between supernaturalism and science? No. Supernaturalism belongs to the pan-human myth-making activity that generates models of personal/cultural coherence and integration through the formulations of supernatural beings and supernatural power. Science belongs to the pan-human analytic activity that generates accurate models to approximate, explain, and use nature.

Is there conflict or cooperation between Christian religion and science? Yes. Some cultural traditions, including the Christian tradition, have attempted to merge supernaturalism and science. Religion is the product of such an attempt, and the debate on cooperation between religion and science is a renewed attempt to subordinate science to supernaturalism. The first special issue of the *Skeptical Inquirer* devoted to science and religion (Frazier 1999) failed to note the fact that religion was a conceptual framework, a cultural category, which the church fathers of the medieval period developed to link science and supernaturalism epistemologically in order to proclaim Christianity as the true explanation of the world and humankind.

Many respected scientists appear to be unaware of this epistemological link. Stephen Jay Gould (1999a) notes that "Science and religion should be equal, mutually respecting partners, each the master of its own domain and with each domain vital to human life in a different way." In his book *Rocks of Ages: Science and Religion in the Fullness of Life* (1999b), Gould writes: "I do not see how science and religion could be unified, or even synthesized, under any common scheme of explanation or analysis; but I also do not understand why the two enterprises should experience conflict. Science tries to document the factual character of the natural world, and to develop theories that coordinate and explain these facts. Religion, on the other hand, operates in the equally important, but utterly different, realm of human purposes, meanings and values—subjects that the factual domain of science might illuminate, but never resolve."

We can agree with Gould's assessment of the relationship (or the lack of relationship) between science and religion *only* if the term *supernaturalism* is substituted for the term *religion*. I am surprised and puzzled that Gould, who has delved into historical issues in many of his essays, failed to make note of the reasons why and how the framework of religion developed.

Arising in the Roman cultural tradition, the Latin term *religio* had multiple meanings such as "the acquisition and possession of supernatural power" and "the performance of rituals for supernatural beings." *Religio* referred to activities that dealt with supernatural powers and beings, and not with a conceptual model of the world. *Religio* was not linked or contrasted with science in the pre-Christian traditions of the West, but the medieval Christian church fathers such as Saint Augustine used *religio* to signify the true knowledge about the nature of the world and humankind. The church, with its hierarchical priesthood, became the custodian of this true knowledge embodied in religion, combining supernaturalism and science.

Attempts to Integrate Religion into Science

Contemporary efforts to represent science and religion as two ways of searching for true knowledge are essentially a continuation and revitalization of the medieval notion of religion. We are inundated with statements such as "evolution is God's way of organizing the natural world," "evolution is God's way of creating human self-awareness," "scientific discoveries reveal God's design," and "science is God's gift to humankind." There are scientists who intentionally or unintentionally confuse the separation of supernaturalism and science by confusing their personal supernaturalism and their objects of inquiry and, in turn, lend scientific legitimacy to religion.

The American scientist Dr. Richard Sneed, in an interview on CNN (1999), advocated human cloning with comments such as the following: God created humans in God's image; God would not have given the intelligence to clone unless God wanted cloning; and cloning was a way of getting close to God. Peter Gosselin (2000) reported that Francis Collins, who runs the Human Genome Research Institute, is a "rare combination of premier scientist and devout Christian. [Collins] professes belief in a God that is beyond the reach of science. He says the pursuit of the genetic code is not, as some worry, an attempt by humans to play God, but only humans' way of admiring God's handiwork. 'God is not threatened by all this' he said in a television interview. 'I think God thinks it's wonderful that we puny creatures are going about the business of trying to understand how our instruction book works, because it's a very elegant instruction book indeed.'"

God can be and is used to justify and legitimize any custom or activity, including science. God can also serve as a vehicle to prevent free inquiry and critical thinking in areas that are prohibited in the name of God, whose prohibition is verified only by the custodians of God and those who accept the custodial claims made in the name of God. Over a hundred years ago, the theologian/biblical scholar/anthropologist William Robertson Smith unsuccessfully defended himself as a scientist who had the moral duty to explore the cultural foundations of Christianity. He was tried for heresy by the Free Church of Scotland and defrocked. His defense was that if God did not want scientific research on discovering the origins of customs, God would not have endowed humans with rationality; he argued that the nonuse of rationality in the furtherance of science was fundamentally a non-Christian attitude. Smith's inquisitors did not accept his defense because in their view the Bible was the revealed truth about the nature of the world and humans, and humans could not fathom the mind of God.

The intellectual history of the past five hundred years has been one of religion attempting to preempt and/or incorporate scientific discoveries as religious truths. Church-affiliated or sectarian universities were built to produce and disseminate religious truths as they were supported by science. If and when scientific discoveries could not be formulated and presented as religious truths, there were inquisitorial persecutions of scientists who were identified as heretics or as atheists. The teaching of "natural theology" and its opposition to the Darwinian model of life-forms prevailed for a long time in academia. We now have departments of religion or religious studies in academia that continue the same intellectual tradition. What occurs today is a much more sophisticated and nuanced attempt to discredit the foundations of science through spurious platitudes such as "religion and science must respect each other," "religion and science must cooperate to seek the basis of reality," "religion and science have a common ground," and "religion and science must seek together to better the conditions of human life." These statements contain expressions that have universal appeal: "respect for each other," "cooperation," and "looking for a common ground for discourse and the search for the betterment of human conditions" are laudable goals of all human beings. But by linking supernaturalism and science epistemologically, a distorted view of science is created, which could lead to the rejection of the scientific method because it discredits God's design or plan.

The scholars of the Enlightenment who did so much to affirm the scientific method and liberate scientific research from supernaturalism failed to recognize that religion was a medieval Christian invention that was developed to oppose what the Church claimed to be pagan, magical, and demonic supernaturalism. Many scholastic treatises on God and the world were viewed by Enlightenment thinkers as facilitating the scientific understanding of the world—for example, formulations concerning a rational God and the rationality of the world, with humans endowed with reason to discover the rationality of the world. Enlightenment thinkers for the most part supported the ethnocentric assumption that religion was superior and more advanced than primitive supernaturalism and that the West had progressed and advanced along the evolutionary ladder because of the applications of rationality to discover the laws of nature and create rational institutions. The Enlightenment, which did so much to revive the Greek ideals of science, was caught in the Christian theological assertions about the nature of religion and supernaturalism. When the exponents of the Enlightenment attacked primitive irrationality as standing in the way of progress, the focus was on non-Western peoples and cultures who, in the view of these scholars, embodied supernaturalism.

The anthropological discourse on humankind was (and is) equally caught in the Christian theological assertions about the nature of religion and supernaturalism. Nineteenth-century sociologists and anthropologists postulated that magic, witchcraft, and divination constituted beliefs and practices that preceded religion and monotheism, and that monotheism and religion manifested themselves only in the higher stages of human mental development. There was also hope that religion would be replaced by science as (and when) the human mind progressed to attain positivistic understanding of natural phenomena. Twentieth-century anthropologists and sociologists devoted considerable time to defining religion, with definitions ranging from religion as supernaturalism to religion as sacred or sanctified values of society, and as the quest for ultimate meaning and reality. Postmodernists have reflected upon whether the definition of religion as supernaturalism is an ethnocentric Western assumption, suggesting that religion should be understood as a system of ordering the world and human life.

Defining Terms and Clarifying Arguments

Perhaps it would clarify discussion if we discarded the use of the term *religion* and substituted the term *supernaturalism*. As I noted earlier, supernaturalism is a cultural universal. Historically humans have created beliefs and practices associated with supernatural beings and supernatural powers, and these beliefs and practices have been used to construct sacred self and group identities and to formulate models or narratives of coherence and meaning to cope with feelings of helplessness, encounters with suffering and injustice, realities of uncertainty, and fear and anxiety associated with sickness and death. Humans have created innumerable forms of the supernatural world with an infinite range of attributes, and this process of creating and maintaining the supernatural world will continue. Science does not attempt to replace or duplicate this creative process, but it attempts to study the relevance and significance of this process in human life.

It is necessary to understand that the concept of religion that developed in the medieval period combines supernaturalism and science to formulate statements about the world and humankind. In this sense, religion does not complement science but preempts and co-opts scientific discourse in support of supernaturalism. The most overt expression of how religion uses science in affirming supernaturalism is found in the evangelical or fundamental Christian perspective known as "scientific creationism" or "creationist science." "Creation scientists" do not see the combination of creation myths and science as an oxymoron but as a way of using the vocabulary of science to foster biblical discourse as scientific.

A recent example of how supernaturalism and science are combined is found in a lawsuit filed in the Minnesota Court of Appeals by a fundamentalist Christian teacher, Lod LaVake. As reported by Joseph Tyrangiel (2000), LaVake's attorney has claimed that "For the first time, we have a teacher who is not asking to teach creationism. He simply wants to teach science the way he thinks—and the way a lot of people think—it should be taught, in a more balanced way." The implication is that Mr. LaVake should be permitted to teach science the way it supports his belief and the belief of many other fundamentalist Christians.

Tyrangiel correctly notes, "Indeed, creationists have become a lot more shrewd. For years they'd propose antievolution laws and lesson plans brimming with religious language, and for years their cases were struck down on constitutional grounds. Like LaVake, they began co-opting the logic of Darwinists and speaking in a softer voice."

Religion does not complement science but preempts and co-opts scientific discourse in support of supernaturalism. We must recognize the importance of supernaturalism in human life and foster its study in terms of why and how humans create and maintain it. The use of the term *religion* to discuss the role of supernaturalism confuses and distorts our understanding of the latter. To the public, the use of *religion* would be more palatable and respectable than the use of the term *supernaturalism* because *supernaturalism* conjures up images of irrational practices such as witchcraft and magic as opposed to religion, which is viewed as a rational, scientific understanding of the world. When scientific research discredits the assumptions of religion, there is conflict between science and religion unless the scientific discoveries are incorporated into the religious framework.

As a first step toward resolving the "science and religion" controversy, and focusing on the real issues that reveal the nature of science and supernaturalism, I suggest that we rename the "departments of religion" in academia and call them "departments of supernaturalism." It is more appropriate to have a discourse on "comparative supernaturalism" than on "comparative religion" because religion, as I noted earlier, is an emic, indigenous category that acquired significance in the medieval period of the Western tradition in an attempt to combine supernaturalism and science within the framework of religion.

References

Barbour, I. G. 1997. *Religion and Science.* New York: HarperCollins.
Easterbrook, G. 1999. "Grappling with Science and Religion." *Los Angeles Times,* March 14.
Estling, R. 2000. "Templeton and the AAAS." *Skeptical Inquirer* 4, no. 24 (July/August).
Frazier, K. 1999. "A Special Issue on Science and Religion." *Skeptical Inquirer* 23, no. 4 (July/August).
Gould, S. J. 1999a. "Dorothy, It's Really Oz." *Time,* August 23.
———. 1999b. *Rocks of Ages: Science and Religion in the Fullness of Life.* New York: W. W. Norton and Company.
Gosselin, P. G. 2000. "Public Project's Chief: Quiet but No Pushover." *Los Angeles Times,* June 27.
Smith, W. C. [1962] 1991. *The Meaning and End of Religion.* Minneapolis: Fortress Press.
Sneed, R. 1999. Interview. Cable News Network, July 11.
Stammer, L. B. 2000. "Physicist Awarded $948,000 Templeton Prize." *Los Angeles Times,* March 23.
Tyrangiel, J. 2000. "The Science of Dissent." *Time,* July 10.

19
SCIENCE VERSUS RELIGION
A Conversation with My Students

BARRY A. PALEVITZ

In the perpetual skirmish between science and religion, biological evolution is a contentious battleground. Despite a series of legal victories for science, which many thought or hoped were the final nails in the coffin of creationism, hostilities still flare. Maybe it's millennialism. Perhaps it reflects unease with modern science and technology. Vigorous antiscience rhetoric coming from the humanities in the guise of postmodernism (see Gross and Levitt 1994; Sokal and Bricmont 1998), some of it antievolution in flavor, may be fueling the fire. Whatever the causes, attacks on evolution haven't ceased. Donald Aguillard, litigant in the landmark 1987 *Edwards* v. *Aguillard* Supreme Court decision on evolution, told the 1998 National Association of Biology Teachers convention in Reno that creationism is thriving in local school districts (see Aguillard 1999). Academia isn't immune either (Berlinski 1996)—here at the University of Georgia, a vocal minority of fundamentalist faculty proclaims the evils of evolution at every opportunity. In response to the controversy, America's premier scientific society recently published yet another forceful statement on the matter (National Academy of Sciences 1998).

Barry Palevitz's "Science and the Versus of Religion: A Conversation with My Students" originally appeared in the *Skeptical Inquirer* 23, no. 4 (July/August 1999).

Because teaching creationism in the public schools is an unconstitutional infringement on the separation of church and state according to the Supreme Court, opponents of evolution no longer approach the issue head-on. One of their tactics is to lower the status of evolution by labeling it "only a theory," a rhetorical trick going back to the 1925 Scopes trial (Larson 1998). Another ploy has been to resurrect the threadbare Argument from Design, which was effectively dismissed on philosophical grounds in the eighteenth century (Hume 1779). But, having lost the "what good is half an eye" battle to developmental biology, creationists have retreated to an equally bogus "intelligent design theory" (Bunk 1998) applied to biochemical pathways and cell structures, most notably in *Darwin's Black Box* (Behe 1996). Of course, for strategic reasons creationists avoid mentioning God as the designer.

After participating in and watching debates with creationists eager to argue about supposed gaps in the evidence for evolution and natural selection, I concluded that arguing about the data is pointless because creationists filter information through a predisposition that has nothing to do with science. It's a matter of what my friends in the humanities call "worldview." Creationists will always see inconsistencies or unexplained phenomena in evolutionary biology that make supernatural intervention an unavoidable conclusion. Science is an easy target in that regard because everything it says is couched in probabilities—certainty isn't in our vocabulary. The public traditionally views science and medicine as authoritative, so uncertainty is easily misinterpreted and manipulated (Toumey 1996). According to increasingly popular mantra, that uncertainty, the fact that we still can't answer all the questions and explain everything about the natural world, by default leads to God (see Johnson 1998b).

The debate, if there is to be one, should instead center on the basic philosophies of science and religion that make them fundamentally different pursuits. In an article in *Discover*, physical anthropologist Matt Cartmill (1998), while lamenting the attacks on evolution by the religious right and now by a new opponent, the academic left, also goes on to chide scientists for ruling God out of the evolutionary equation. To do so is a personal, not a scientific, statement, he says. I think Cartmill is mistaken. Creationists want to redefine science in the public eye in order to accommodate their religious views. Since creationism has failed to gain credibility as science according to commonly accepted norms, proponents want to change the rules by altering the public's perception of the nature of science. By calling the scientific concept of purposeless evolution "an article of faith," and by insisting that Stephen Jay Gould's ideas are "a profession of his religious beliefs," Cartmill's position plays right into their hands.

An incident in my honors botany class helps focus the relationship between science and religion. A genetics faculty member delivered a lecture on God and genes. Because genes explain so much of what used to be God's turf, he argued, we should reexamine religion in that light (see Avise 1998). Knowing the lecture would be thought provoking, I asked my students to attend and write a brief essay voicing their opinions. Many students wrote that the speaker had wrongly excluded religion from the realm of science. I was so surprised that I spent the next class period revisiting the subject.

Religion in Science

As we sat outside on a pleasant afternoon, I fielded questions from the students about the place of religion in science. Some of the ground we covered was familiar to them, but a lot of it wasn't. I first floated the standard argument that science and religion see the world in very different ways and use contrasting methods and standards. Science offers material explanations based on objective observation and measurement; if it accepts the supernatural, troubling questions arise. For example, which of the world's many creation stories qualifies as the official alternative to evolution? In the United States, the public would vote for Genesis, but do we prefer the literal interpretation espoused by the fundamentalists or one with a more allegorical spin? Another problem: If we add religion to the mix, why not astrology or other nonscientific viewpoints? My students weren't satisfied.

As we talked, I tried to get them to more fully understand the holy grail of data in science and the concept that understanding may arrive with the next observation or measurement. That's why we spend ceaseless hours and dollars pursuing new ways to sequence genes and view distant galaxies. That's also why supernatural explanations are unnecessary and counterproductive. If the answer isn't obvious because of insufficient data, we wait for the data to appear in the literature or do the necessary experiments ourselves. We don't punt in favor of the supernatural, as Michael Behe did. History clearly shows that with enough time and data, the unexplained is demystified. No natural phenomenon, not one, has ever been shown to have a supernatural cause based on objective, material evidence. On the contrary, fear, superstition, and myth surrounding everything from lightning to disease have consistently fallen to rational inquiry. As for all the unanswered questions, that's what science is about! Old questions are answered (or old answers modified) as new ones arise. The ex-editor of *Nature*, John Maddox, put it well: "As in the past, deepened understanding will provoke questions we do not yet have the wit to ask" (Maddox 1998).

Creationists have consistently made us defend evolution when in fact the burden of proof is on them. Evolution is the accepted scientific explanation for how life arrived on the scene, and it's supported by a mountain of data. If creationists believe their ideas are correct, it is they who have to make a convincing case. Because the name of the game is science, we should force them to play by science's rules and demand that they defend their position based on positive scientific evidence; supposedly unexplainable phenomena and attacks on evolution aren't enough. Behe's book attracted a lot of media attention (and sales) despite the absence of any positive evidence in favor of supernatural design. But then again, is that surprising? Ultimately, the supernatural cannot be tested by the material methods of science, and we should force creationists to admit it.

One of my students asked if I "believe in" evolution. Instead, I pointed out a problem with his choice of words. Scientists do not believe in evolution; they believe it. That's not just a matter of semantics but a real distinction. To believe in evolution implies faith, like believing in fairies. Scientists believe evolution because overwhelming data support it, and they will do so until data say otherwise. Creationists like to argue that scientists accept evolution in the believe-in mode, i.e., as a matter of faith. Those who make that argument confuse and mislead. Darwin published his ideas on the basis of his own and previous observations. And while mechanisms are hotly debated, the data since then support evolution; it's not a matter of faith.

Compromising Science

My students were still troubled—thinking that there must be room for religion *somewhere* in science. Surely a little religion here and there wouldn't hurt. I tried to explain that there's no such thing as just a little religion in science. Open a crack in the evidentiary standards of science and somebody will drive a tank through it. Open it a little, and a flood of vested interests will line up for equal time.

But surely you can *compromise*, Professor Palevitz? After all, isn't that *fair*? My response was to consider the problem that proposition raises: How should I compromise? Which bit of scientific turf should I concede? Which scintilla of objective evidence should I ignore in favor of the subjective and untestable? After all, we're not talking about horse trading over congressional appropriations where swapping a dam for an F-16 is the norm. No, this kind of compromise is much more difficult, comparable to the Sophie's choice dilemma of a parent forced to choose between children. It's an impossible choice for the parent, and equally impossible for the scientist.

Some creationists are willing to admit that Earth is really as old as science shows. Some even concede that other species may have evolved, but God stepped in for *Homo sapiens*. Isn't that good middle ground, professor? The answer is no. If we accept supernatural intervention for humans, why not for other organisms? After all, there is no scientific justification for removing humanity from the rest of biology. If we accept a little of the subjective, why not go whole hog and admit that Earth may indeed be only 6,000 years old and fossils are a divine trick to test our faith, as some creationists hold? Creationism comes in various flavors, and there is no clear line separating the somewhat palatable from the just plain awful. The only choice is to stick to objective scientific evidence.

The basic flaw in Cartmill's position in *Discover* is that science isn't free to yield the inch of ground he calls for. Science has no basis for acknowledging the possible influence of God in evolution, no matter how small or hypothetical. It's not that science excludes God so much as it has no way of dealing with the concept. Since it cannot be approached by scientific principles and methods, the supernatural is automatically off limits as an explanation of the natural world. It's not a factor in the equation; it's not in the same ballpark. All science can go on is material evidence, which says that the supernatural is not necessary for explaining biological diversity. We can acknowledge supernatural beliefs about life, but those beliefs are irrelevant scientifically.

A bit more should be said about the word "fair" since it's often trotted out to convince people that creationism should be taught side by side with evolution. After all, what can be wrong with giving students both sides of the story—it's democratic; it's the American way. *But it's not that simple.* Again, there are many religious traditions, not one. Which do we anoint as the official version, or do we discuss all of them? At a more fundamental level, is science really about fairness and democracy? Hardly. True, science is a free marketplace of ideas and many people have a voice in that marketplace. Scientists debate alternative hypotheses, and often, in the absence of a clear winner, some form of agreement emerges as a basis for new experiments. But there the similarity to a New England town meeting ends, because an idea, consensus notwithstanding, lasts only as long as data support it. Fairness has nothing to do with it. Religion doesn't make the grade because it offers no data. It simply isn't science.

The fairness argument is also hypocritical. Precollege students study science perhaps one hour each school day, and only a tiny percentage of the time is devoted to evolution. In contrast, they spend about seventeen hours a day and all weekend outside of school, free to hear the "alternative." Yet creationists covet even the small amount of time in biology class, in the name of fairness.

To me, the differences between science and religion are stark, but many people miss the distinction. Perhaps we are not doing a good enough job explaining the nature of scientific inquiry and rational problem solving to our students. We teach facts and principles in our science classes, but apparently not enough about the philosophical underpinnings of the scientific process and the way scientists view the natural world. The little philosophy we do teach often is a stale version of the "scientific method" that only vaguely resembles the way things really happen. We confer on our graduate students the degree of Doctor of Philosophy, yet what kind of philosophical training do we provide? Many of them will become teachers or otherwise interpret science for the public, but can they explain how science works?

A recent survey of Pennsylvania high school teachers (Osif 1997) showed that nearly 40 percent agreed that creationism should be taught in the public schools, a figure consistent with data from other states (also see Aguillard 1999). Surprisingly, more than 20 percent of the Pennsylvania science teachers did not accept the principle that evolution is central to the study of biology. According to the survey's author, "This inability or refusal to judge information on its scientific validity is chilling." Opinions about evolution did not depend on the teachers' academic training. In other words, learning facts and concepts wasn't enough. Instead, Osif suggested that "study of the nature, rules, and philosophy of science may be one step in addressing this issue," a conclusion emphasized by the National Academy of Sciences in its 1998 report.

Perhaps the most pernicious message of creationists is this: If only we deny it, evolution will go away. If enough people are convinced, nature will follow suit. The radical postmodernists have a similar fantasy: If their objections to biological determinism (essentialism) are loud enough, perhaps genetic influences on human behavior will disappear (Rosser 1986; Nelkin and Lindee 1995).

This kind of message should send shivers down the spine of anybody interested in honest, objective inquiry. Science discovers the principles upon which the universe operates, it doesn't construct them, and it's not free to ignore them. Nobody, scientists included, has the ability to legislate how the natural world operates based on religious or political belief. One cannot simply insist that the emperor is wearing a fine set of clothes and expect it to be true. We cannot substitute how we would like the world to be for how it actually is.

Differing Spheres

Unfortunately, scientists themselves often fray the boundaries between science and religion, for example, by referring to "God particles" or the mind of God. In an effort to bridge the gap between science and religion, some claim that science supports Scripture and can even summon proof of God's existence (Schroeder 1997; Begley 1998; Gibbs 1998; Johnson 1998a). Others embrace the anthropic principle, a quasi-supernatural belief that the universe was created just so with us in mind (Glynn 1997). In an effort to seem evenhanded or fair when discussing evolution, some of my colleagues tell students that religion and science are equally worthwhile ways of pursuing knowledge and truth. Cartmill's call for compromise and humility by scientists is another side of the same coin. But is that approach really useful? I think not. We should more correctly point out that religion and science are equally valid *at what they do individually, but not in each other's sphere.* Science is an excellent system for explaining the natural world, as history has amply shown, but it says nothing about morality. Conversely, religion may be a useful framework for moral guidance, but it is a poor system for systematically investigating the natural world. To maintain that Genesis (or other creation accounts) and science are equally valid in explaining how life arrived at its present state only muddies the waters. Different, yes; equal, no.

As the sun waned that afternoon, a final misunderstanding surfaced. A student wanted to know why scientists are so spirituality bereft. Shocked, I responded by distinguishing between spirituality and religion. It is religion that has no place in science; spirituality is not only possible but even advantageous if it reinforces love of the natural world and motivation to find out how it works. I consider myself a very spiritual person: I am awed every time I snorkel on a coral reef, humbled by a crisp December night full of stars, forever amazed by a dividing cell, and inspired by human creativity. Like physicist Richard

Feynman (see Sykes 1994), my appreciation of nature's beauty is magnified by knowledge of its mechanisms. I resent the implication that scientists have no spiritual dimension, and if the war being waged against science by segments of the religious and academic communities helps spread that myth, scientists have even more reason to resist it.

References

Aguillard, Donald W. 1999. "Evolution Education in Louisiana Public Schools: A Decade Following *Edwards* v. *Aguillard.*" *American Biology Teacher*, in press.

Avise, John C. 1998. *The Genetic Gods*. Cambridge: Harvard University Press.

Begley, Sharon. 1998. "Science Finds God." *Newsweek* 132, no. 3: 46–51.

Behe, Michael J. 1996. *Darwin's Black Box: The Biochemical Challenge to Evolution*. New York: Free Press.

Berlinski, David. 1996. "The Deniable Darwin." *Commentary* 101, no. 6: 19–29.

Bunk, Steve. 1998. "In a Darwinian World, What Chance for Design?" *Scientist* 12, no. 8: 4.

Cartmill, Matt. 1998. "Oppressed by Evolution." *Discover* 19, no. 3: 78–83.

Gibbs, W. Wayt. 1998. "Beyond Physics: Renowned Scientists Contemplate the Evidence for God." *Scientific American* 279, no. 2: 20–22.

Glynn, Patrick. 1997. *God: The Evidence*. Rocklin, Calif.: Prima Publishing.

Gross, Paul R., and Norman Levitt. 1994. *Higher Superstition*. Baltimore: Johns Hopkins University Press.

Hume, David. 1779. *Dialogues Concerning Natural Religion*. Edited and annotated in 1948 by Norman K. Smith. New York: Social Sciences Publishers.

Johnson, George. 1998a. "Science and Religion: Bridging the Great Divide." *New York Times*, June 30, p. F4.

Johnson, George. 1998b. "Science and Religion Cross Their Line in the Sand." *New York Times*, July 12, sec. 4:1.

Larson, Edward J. 1997. *Summer for the Gods: The Scopes Trial and America's Continuing Debate over Science and Religion*. New York: Basic Books.

Maddox, John. 1998. "Nobody Can Tell What Lies Ahead." *New York Times*, November 10, p. D5.

National Academy of Sciences. 1998. *Teaching About Evolution and the Nature of Science*. Washington, D.C.: National Academy Press.

Nelkin, Dorothy, and M. Susan Lindee. 1995. *The DNA Mystique: The Gene as a Cultural Icon*. New York: W. H. Freeman.

Osif, Bonnie A. 1997. "Evolution and Religious Beliefs: A Survey of Pennsylvania High School Teachers." *American Biology Teacher* 59: 552–56.

Rosser, Sue V. 1986. *Teaching Science and Health from a Feminist Perspective.* New York: Pergamon Press.

Schroeder, Gerald L. 1997. *The Science of God.* New York: Free Press.

Sokal, Alan, and Jean Bricmont. 1998. *Fashionable Nonsense: Postmodern Philosophers' Abuse of Science.* New York: St. Martin's Press.

Sykes, Christopher. 1994. *No Ordinary Genius: The Illustrated Richard Feynman.* New York: W. W. Norton.

Toumey, Christopher P. 1996. *Conjuring Science.* New Brunswick: Rutgers University Press.

20
CREDO

ARTHUR C. CLARKE

For thousands of years the subtlest minds of the human species have been focused on the great questions of life and death, of time and space—and of man's place in the universe. The answers have been encapsulated in the holy books of countless religions and whole libraries of philosophy, folklore, and myth.

Can our age contribute anything both new and true to these ancient debates? I believe so. We have been lucky enough to live at a time when knowledge that once seemed forever beyond reach can be found in elementary schoolbooks. Our generation has seen the far side of the Moon, and close-ups of all the major bodies circling the Sun. We have opened the Pandora's box of the atomic nucleus. And perhaps most marvelous of all, we have uncovered the secret of life itself, in the endless twining and untwining of the DNA spiral. This is perhaps the greatest discovery in the whole history of science, yet even now it is barely thirty years old.

Arthur C. Clarke's "Credo" originally appeared in Clifton Fadiman, ed., *Living Philosophies* (New York: Doubleday, 1991). Copyright © 1990 by Clifton Fadiman. Used by permission of Doubleday, a division of Random House, Inc. It was republished in *Greetings, Carbon-Based Life Bipeds! Collected Essays 1934–1998* (New York: St. Martin's Press, 1999) and again in the *Skeptical Inquirer* 25, no. 5 (September/October 2001).

There are those who claim not to be impressed by such achievements, arguing that science deals with unimportant questions that can be solved, while religion is concerned with important ones that can't. The logical positivists would maintain that this is nonsense; if a problem can't be solved, at least in principle, it doesn't really exist. In other words, there's no such animal as metaphysics.

Without knowing it, I became a logical positivist at about the age of ten. Every Sunday, I was supposed to make the two-mile walk to the local Church of England—it was a long time before I discovered there was any other variety—to attend a service for the village youth. To encourage us to sit through the sermons, we were rewarded with stamps illustrating scenes from the Bible. When we had filled an album with these, we were entitled to an "outing"— i.e., a bus trip to some exotic and remote part of Somerset, perhaps as far as twenty miles away. I stuck with it for a few weeks, then decided—to quote Churchill's famous memorandum on the necessity of ending sentences with a preposition—"This is nonsense up with which I will not put."

Half a century of travel, reading, and contact with other faiths has endorsed that early insight.

Now I myself am not completely innocent, according to one of the last letters I received from the great biologist J. B. S. Haldane. Shortly before he died (going not gently but heroically into the good night with a witty poem entitled "Cancer Can Be Fun") he wrote: "I would like to see you awarded a prize for theology, as you are one of the very few living persons who has written anything original about God. You have in fact, written several mutually incompatible things. . . . If you had stuck to one theological hypothesis you might be a serious public danger."

I am only sorry that JBS never had a chance to criticize my later (doubtless yet more incompatible) speculations, developed in the novels *The Fountains of Paradise* and *The Songs of Distant Earth*. He would, I am sure, have enjoyed this specimen from *Fountains*:

> There can be no such subject as comparative religion as long as we study only the religions of man. . . . If we find that religion occurs exclusively among intelligent analogs of apes, dolphins, elephants, dogs, etc., but not among extraterrestrial computers, termites, fish, turtles, or social amoebae, we may have to draw some painful conclusions. . . . Perhaps both love and religion can arise only among mammals, and for

much the same reasons. This is also suggested by a study of their pathologies; anyone who doubts the connection between religious fanaticism and perversion should take a long, hard look at the *Malleus Maleficarum* or Huxley's *The Devils of Loudun.*

But I am quite serious about the profound philosophical importance of the Search for Extra-Terrestrial Intelligence (SETI); this may be its supreme justification. The fact that we have not yet found evidence for life—much less intelligence—beyond this Earth does not surprise or disappoint me in the least. Our technology must still be laughably primitive; we may well be like jungle savages listening for throbbing of tom-toms, while the ether around them carries more words per second than they could utter in a lifetime.

The greatest tragedy in mankind's entire history may be the hijacking of morality by religion. However valuable—even necessary—that may have been in enforcing good behavior on primitive peoples, their association is now counterproductive. Yet at the very moment when they should be decoupled, sanctimonious nitwits are calling for a return to morals based on superstition.

Having disposed of religion (at least until next Wednesday), let us consider something really important: God—aka Allah/Brahma/Jehovah, etc. ad infinitum. In *The Songs of Distant Earth*, I distinguished between two aspects of this hypothetical entity, calling them Alpha and Omega to defuse emotional reactions.

Alpha might be identified with the jealous God of the Old Testament, who watches over all creatures ("His eye is on the sparrow") and rewards good and evil in some vaguely described afterlife. Even today, belief in Alpha is fading fast; I suggested that early in the next millennium the rise of "statistical theology" would prove that there is no supernatural intervention in human affairs. Nor does the "problem of evil" exist; it is an inevitable consequence of the bell-shaped curve of normal distribution.

Unfortunately, most people do not understand even the basic elements of statistics and probability, which is why astrologers and advertising agencies flourish. If you want to start an interesting fight, say in a loud voice at your next cocktail party, "Fifty percent of Americans (or whatever) are mentally subnormal." Then watch all those annoyed by this mathematical tautology instantly pigeonhole themselves.

I also, rather mischievously, demolished Alpha by invoking the ghost of Kurt Gödel, whose notorious "incompleteness of knowledge" theorem quite obviously rules out the existence of an omniscient being. However, this is an area where logic gets you nowhere. Belief—or disbelief—in Alpha appears to be irrevocably programmed into most people at an early age.

A man I admire, who has held the highest medical position in the United States, recently declared, "There are no atheists at the bedside of a dying child." It is a compassionate statement, nobly expressed, with which every humane person must sympathize. But, with all respect, it is simply untrue.

Nor have I ever felt a need for Alpha on the several occasions when I thought I was about to die (in each case, at a depth of embarrassingly few fathoms). Certainly the notion of appealing for divine help never entered my mind; I was much too busy thinking, "How do I get out of this ridiculous situation?"

Omega—the Creator of Everything—is a much more interesting character than Alpha, and not so easily dismissed. Although irredeemable agnostics may smile at Edward Young's "The undevout astronomer is mad," no intelligent persons can contemplate the night sky without a sense of awe. The mind-boggling vista of exploding supernovae and hurtling galaxies does seem to require a certain amount of explaining: to answer the question "Why is the universe here?" with the retort "Where else would it be?" is somehow not very satisfying. Although—the logical positivists would be pleased—it may be all the answer that is needed, because the question itself may not make sense.

Let me offer an analogy, suggested by a conversation I once had with C. S. Lewis. We science-fiction authors are always picking each other's brain, and Lewis asked me what the horizon would look like (ignoring atmospheric absorption) on a really enormous planet—one not thousands, but millions, of kilometers in radius.

Any inhabitants would be convinced that they were living on a perfectly flat plane and might fight holy wars over the rival doctrines (a) the world goes on forever and ever; (b) you'll fall off when you reach the edge. But to us, there is no problem. We have watched the globe of the Earth floating on our television screens and have no difficulty in understanding why both flatlander cults are wrong. If they ever got around to making spaceships, their religious disputations would be ended.

So it is very, very risky to maintain that, as the old B-grade movies loved to intone, "There is some knowledge not meant for Man." I am fond of quoting a monumental gaffe made by Auguste Comte, who told the astronomers in no uncertain terms just what they could ever expect to know about other worlds—"We may determine their forms, their distances, their bulk, their motions—but we can never know anything of their chemical or mineralogical structure; and, much less, that of organized beings living on their surface."

Within a century of Comte's death, thanks to the invention of the spectroscope, much of astronomy had become astrochemistry—a science he had roundly declared impossible. I wonder what he would have said about space exploration, had anyone been rash enough to suggest such an absurdity to him.

So it may be that questions which now seem almost beyond conjecture may one day be conclusively settled. The limits of space, the beginning and ending of time, the origin of matter and energy, may have no mysteries to our remote descendants. And many of the questions we ask of the universe may turn out to be completely meaningless—as certain theories on the frontiers of modern physics tantalizingly suggest.

I felt this very strongly when I was privileged to make a television program, modestly entitled "God, the Universe and Everything Else" with Newton's successor Dr. Stephen Hawking. If you have not read *A Brief History of Time*, please rectify the omission—and read the bits about "imaginary time." Thank you; that saves me a lot of hand waving, trying to explain how our own views of past and future may be as naive as the flatlanders' ideas about the geometry of their giant planet.

The extraordinary success of Dr. Hawking's book is one of the best pieces of news from the popular science—indeed, educational—front for many years. I have been appalled by the way in which the United States (and much of the world, East and West) appears to be sinking into cultural barbarism, harangued by the fundamentalist ayatollahs of the airwaves, its bookstores, and newsstands poisoned with mind-rotting rubbish about astrology, UFOs, reincarnation, ESP, spoon-bending, and especially "creationism." This last—which implies that the marvelous and inspiring story of evolution, so clearly recorded in the geological strata, is all a cosmic practical joke—helps me to understand the revulsion that a devout Muslim must feel toward *The Satanic Verses*. If there is indeed such a thing as blasphemy, it is here. . . .

The Pontifical Academy of Science—which I have been honored to address—has now firmly stated: "Masses of evidence render the application of the concept of evolution to man and the other primates beyond serious dispute."

I began this essay by saying that men have debated the problems of existence for thousands of years—and that is precisely why I am skeptical about most of the answers. One of the great lessons of modern science is that millennia are only moments. It is not likely that ultimate questions will be settled in such short periods of time, or that we will really know much about the universe while we are still crawling around in the playpen of the solar system.

So let us recognize that there is much concerning which we must reserve judgment, and refuse to take seriously all dogmas and revelations whose acceptance demands faith. They have been proved wrong countless times in the past; they will be proved wrong again in the ages to come.

And worse than wrong. Who can forget Jacob Bronowski, in his superb television series, *The Ascent of Man*, standing among the ashes of his relatives at the Auschwitz crematorium and reminding us: "This is how men behave when they believe they have absolute knowledge." This is how they are still behaving—in Ireland, in Lebanon, in Palestine, in Iran—and at this very moment, alas, in my own Sri Lanka.

Yet, if absolute knowledge is unattainable, someday most of the great truths may be established—if not with absolute certainty, then beyond all reasonable doubt. Do not be impatient; there is plenty of time.

How much time, we are only now beginning to appreciate. In a famous essay, "Time without End," Freeman Dyson speculated that a high-technology cosmic intelligence might even be able to make itself, quite literally, immortal.

So let me end with the final chapter, "The Long Twilight," from my *Profiles of the Future: An Inquiry into the Limits of the Possible.*

> Whether Freeman Dyson's vision (some would say nightmare) of eternity is true or not, one thing seems certain. Our galaxy is now in the brief springtime of its life—a springtime made glorious by such brilliant blue-white stars as Vega and Sirius, and, on a more humble scale, our own Sun. Not until all these have flamed through their incandescent youth, in a few fleeting billions of years, will the real history of the universe begin.
>
> It will be a history illuminated only by the reds and infrareds of dully glowing stars that would be almost invisible to our eyes; yet the somber hues of that all-but-eternal universe may be full of color and beauty to whatever strange beings have adapted to it. They will know that before them lie, not the millions of years in which we measure eras of geology, nor the billions of years which span the past lives of the stars, but years to be counted literally in trillions.

They will have time enough, in those endless aeons, to attempt all things, and to gather all knowledge. They will be like gods, because no gods imagined by our minds have ever possessed the powers they will command. But for all that, they may envy us, basking in the bright afterglow of Creation; for we knew the universe when it was young.

IV

Science and Ethics
Two Magisteria

21
NONOVERLAPPING MAGISTERIA

STEPHEN JAY GOULD

Incongruous places often inspire anomalous stories. In early 1984, I spent several nights at the Vatican housed in a hotel built for itinerant priests. While
pondering over such puzzling issues as the intended function of the bidet in
each bathroom, and hungering for something more than plum jam on my
breakfast rolls (why did the basket only contain hundreds of identical plum
packets and not a one of, say, strawberry?), I encountered yet another among
the innumerable issues of contrasting cultures that can make life so expansive
and interesting. Our crowd (present in Rome to attend a meeting on nuclear
winter, sponsored by the Pontifical Academy of Sciences) shared the hotel with
a group of French and Italian Jesuit priests who were also professional scientists. One day at lunch, the priests called me over to their table to pose a problem that had been troubling them. What, they wanted to know, was going on
in America with all this talk about "scientific creationism"? One of the priests
asked me: "Is evolution really in some kind of trouble; and, if so, what could

Stephen Jay Gould's "Nonoverlapping Magisteria" is excerpted from his book
Leonardo's Mountain of Clams and the Diet of Worms (New York: Three Rivers Press,
1998). Copyright © 1998 by Turbo, Inc. Used by permission of Harmony Books, a
division of Random House, Inc. It later appeared in the *Skeptical Inquirer* 23, no. 4
(July/August 1999).

such trouble be? I have always been taught that no doctrinal conflict exists between evolution and Catholic faith, and the evidence for evolution seems both utterly satisfying and entirely overwhelming. Have I missed something?"

A lively pastiche of French, Italian, and English conversation then ensued for half an hour or so, but the priests all seemed reassured by my general answer—"Evolution has encountered no intellectual trouble; no new arguments have been offered. Creationism is a home-grown phenomenon of American sociocultural history—a splinter movement (unfortunately rather more of a beam these days) of Protestant fundamentalists who believe that every word of the Bible must be literally true, whatever such a claim might mean." We all left satisfied, but I certainly felt bemused by the anomaly of my role as a Jewish agnostic, trying to reassure a group of priests that evolution remained both true and entirely consistent with religious belief.

Another story in the same mold: I am often asked whether I ever encounter creationism as a live issue among my Harvard undergraduate students. I reply that only once, in thirty years of teaching, did I experience such an incident. A very sincere and serious freshman student came to my office with a question that had clearly been troubling him deeply. He said to me, "I am a devout Christian and have never had any reason to doubt evolution, an idea that seems both exciting and well documented. But my roommate, a proselytizing evangelical, has been insisting with enormous vigor that I cannot be both a real Christian and an evolutionist. So tell me, can a person believe both in God and in evolution?" Again, I gulped hard, did my intellectual duty, and reassured him that evolution was both true and entirely compatible with Christian belief—a position that I hold sincerely, but still an odd situation for a Jewish agnostic.

These two stories illustrate a cardinal point, frequently unrecognized but absolutely central to any understanding of the status and impact of the politically potent, fundamentalist doctrine known by its self-proclaimed oxymoron as "scientific creationism"—the claim that the Bible is literally true, that all organisms were created during six days of twenty-four hours, that the Earth is only a few thousand years old, and that evolution must therefore be false. Creationism does not pit science against religion (as my opening stories indicate), for no such conflict exists. Creationism does not raise any unsettled intellectual issues about the nature of biology or the history of life. Creationism is a local and parochial movement, powerful only in the United States among

Western nations, and prevalent only among the few sectors of American Protestantism that choose to read the Bible as an inerrant document, literally true in every jot and tittle.

I do not doubt that one could find an occasional nun who would prefer to teach creationism in her parochial school biology class, or an occasional rabbi who does the same in his yeshiva, but creationism based on biblical literalism makes little sense either to Catholics or Jews, for neither religion maintains any extensive tradition for reading the Bible as literal truth, other than illuminating literature based partly on metaphor and allegory (essential components of all good writing), and demanding interpretation for proper understanding. Most Protestant groups, of course, take the same position—the fundamentalist fringe notwithstanding.

The argument that I have just outlined by personal stories and general statements represents the standard attitude of all major Western religions (and of Western science) today. (I cannot, through ignorance, speak of Eastern religions, though I suspect that the same position would prevail in most cases.) The *lack of conflict* between science and religion arises from a *lack of overlap* between their respective domains of professional expertise—science in the empirical constitution of the universe, and religion in the search for proper ethical values and the spiritual meaning of our lives. The attainment of wisdom in a full life requires extensive attention to both domains—for a great book tells us both that the truth can make us free, and that we will live in optimal harmony with our fellows when we learn to do justly, love mercy, and walk humbly.

In the context of this "standard" position, I was enormously puzzled by a statement issued by Pope John Paul II on October 22, 1996, to the Pontifical Academy of Sciences, the same body that had sponsored my earlier trip to the Vatican. In this document, titled "Truth Cannot Contradict Truth," the pope defended both the evidence for evolution and the consistency of the theory with Catholic religious doctrine. Newspapers throughout the world responded with front-page headlines, as in the *New York Times* for October 25: "Pope Bolsters Church's Support for Scientific View of Evolution."

Now I know about "slow news days," and I do allow that nothing else was strongly competing for headlines at that particular moment. Still, I couldn't help feeling immensely puzzled by all the attention paid to the pope's statement (while being wryly pleased, of course, for we need all the good press we can get, especially from respected outside sources). The Catholic Church does not oppose evolution, and has no reason to do so. Why had the pope issued such a statement at all? And why had the press responded with an orgy of worldwide front-page coverage?

I could only conclude at first, and wrongly as I soon learned, that journalists throughout the world must deeply misunderstand the relationship between science and religion, and must therefore be elevating a minor papal comment to unwarranted notice. Perhaps most people really do think that a war exists between science and religion, and that evolution cannot be squared with a belief in God. In such a context, a papal admission of evolution's legitimate status might be regarded as major news indeed—a sort of modern equivalent for a story that never happened, but would have made the biggest journalistic splash of 1640: Pope Urban VIII releases his most famous prisoner from house arrest and humbly apologizes: "Sorry, Signor Galileo . . . the Sun, er, is central."

But I then discovered that such prominent coverage of papal satisfaction with evolution had not been an error of non-Catholic anglophone journalists. The Vatican itself had issued the statement as a major news release. And Italian newspapers had featured, if anything, even bigger headlines and longer stories. The conservative *Il Giornale*, for example, shouted from its masthead: "Pope Says We May Descend from Monkeys."

Clearly, I was out to lunch; something novel or surprising must lurk within the papal statement, but what could be causing all the fuss?—especially given the accuracy of my primary impression (as I later verified) that the Catholic Church values scientific study, views science as no threat to religion in general or Catholic doctrine in particular, and has long accepted both the legitimacy of evolution as a field of study and the potential harmony of evolutionary conclusions with Catholic faith.

As a former constituent of Tip O'Neill, I certainly know that "all politics is local"—and that the Vatican undoubtedly has its own internal reasons, quite opaque to me, for announcing papal support of evolution in a major statement. Still, I reasoned that I must be missing some important key, and I felt quite frustrated. I then remembered the primary rule of intellectual life: When puzzled, it never hurts to read the primary documents—a rather simple and self-evident principle that has, nonetheless, completely disappeared from large sectors of the American experience.

I knew that Pope Pius XII (not one of my favorite figures in twentieth-century history, to say the least) had made the primary statement in a 1950 encyclical entitled *Humani Generis*. I knew the main thrust of his message: Catholics could believe whatever science determined about the evolution of the human body, so long as they accepted that, at some time of his choosing, God had infused the soul into such a creature. I also knew that I had no problem with this argument—for, whatever my private beliefs about souls, science cannot touch such a subject and therefore cannot be threatened by any theological position on such a legitimately and intrinsically religious issue. Pope Pius XII, in other words, had properly acknowledged and respected the separate domains of science and theology. Thus, I found myself in total agreement with *Humani Generis*—but I had never read the document in full (not much of an impediment to stating an opinion these days).

I quickly got the relevant writings from, of all places, the Internet. (The pope is prominently on line, but a luddite like me is not. So I got a cyberwise associate to dredge up the documents. I do love the fracture of stereotypes implied by finding religion so hep and a scientist so square.) Having now read in full both Pope Pius's *Humani Generis* of 1950 and Pope John Paul's proclamation of October 1996, I finally understand why the recent statement seems so new, revealing, and worthy of all those headlines. And the message could not be more welcome for evolutionists, and friends of both science and religion.

The text of *Humani Generis* focuses on the *Magisterium* (or Teaching Authority) of the Church—a word derived not from any concept of majesty or unquestionable awe, but from the different notion of teaching, for *magister* means "teacher" in Latin. We may, I think, adopt this word and concept to express the central point of this essay and the principled resolution of supposed "conflict" or "warfare" between science and religion. No such conflict should exist because each subject has a legitimate magisterium, or domain of teaching authority—and these magisteria do not overlap (the principle that I would like to designate as NOMA, or "nonoverlapping magisteria"). The net of science covers the empirical realm: what is the universe made of (fact) and why does it work this way (theory). The net of religion extends over questions of moral meaning and value. These two magisteria do not overlap, nor do they encompass all inquiry (consider, for starters, the magisterium of art and the meaning of beauty). To cite the usual clichés, we get the age of rocks, and religion retains the rock of ages; we study how the heavens go, and they determine how to go to heaven.

This resolution might remain entirely neat and clean if the nonoverlapping magisteria of science and religion stood far apart, separated by an extensive no-man's-land. But, in fact, the two magisteria bump right up against each other, interdigitating in wondrously complex ways along their joint border. Many of our deepest questions call upon aspects of both magisteria for different parts of a full answer—and the sorting of legitimate domains can become quite complex and difficult. To cite just two broad questions involving both evolutionary facts and moral arguments: Since evolution made us the only earthly creatures with advanced consciousness, what responsibilities are so entailed for our relations with other species? What do our genealogical ties with other organisms imply about the meaning of human life?

Pius XII's *Humani Generis* (1950), a highly traditionalist document written by a deeply conservative man, faces all the "isms" and cynicisms that rode the wake of World War II and informed the struggle to rebuild human decency from the ashes of the Holocaust. The encyclical bears the subtitle "concerning some false opinions which threaten to undermine the foundations of Catholic doctrine," and begins with a statement of embattlement:

> Disagreement and error among men on moral and religious matters have always been a cause of profound sorrow to all good men, but above all to the true and loyal sons of the Church, especially today, when we see the principles of Christian culture being attacked on all sides.

Pius lashes out, in turn, at various external enemies of the Church: pantheism, existentialism, dialectical materialism, historicism, and, of course and preeminently, communism. He then notes with sadness that some well-meaning folks within the Church have fallen into a dangerous relativism—"a theological pacifism and egalitarianism, in which all points of view become equally valid"—in order to include those who yearn for the embrace of Christian religion, but do not wish to accept the particularly Catholic magisterium.

Speaking as a conservative's conservative, Pius laments:

> Novelties of this kind have already borne their deadly fruit in almost all branches of theology. . . . Some question whether angels are personal beings, and whether matter and spirit differ essentially. . . . Some even say that the doctrine of Transubstantiation, based on an antiquated philosophic notion of substance, should be so modified that the Real Presence of Christ in the Holy Eucharist be reduced to a kind of symbolism.

Pius first mentions evolution to decry a misuse by overextension among zealous supporters of the anathematized "isms":

> Some imprudently and indiscreetly hold that evolution . . . explains the origin of all things. . . . Communists gladly subscribe to this opinion so that, when the souls of men have been deprived of every idea of a personal God, they may the more efficaciously defend and propagate their dialectical materialism.

Pius presents his major statement on evolution near the end of the encyclical, in paragraphs 35 through 37. He accepts the standard model of nonoverlapping magisteria (NOMA) and begins by acknowledging that evolution lies in a difficult area where the domains press hard against each other. "It remains for Us now to speak about those questions which, although they pertain to the positive sciences, are nevertheless more or less connected with the truths of the Christian faith."[1]

Pius then writes the well-known words that permit Catholics to entertain the evolution of the human body (a factual issue under the magisterium of science), so long as they accept the divine creation and infusion of the soul (a theological notion under the magisterium of religion).

> The Teaching Authority of the Church does not forbid that, in conformity with the present state of human sciences and sacred theology, research and discussions, on the part of men experienced in both fields, take place with regard to the doctrine of evolution, in as far as it inquires into the origin of the human body as coming from pre-existent and living matter—for the Catholic faith obliges us to hold that souls are immediately created by God.

I had, up to here, found nothing surprising in *Humani Generis*, and nothing to relieve my puzzlement about the novelty of Pope John Paul's recent statement. But I read further and realized that Pius had said more about evolution, something I had never seen quoted, and something that made John Paul's statement most interesting indeed. In short, Pius forcefully proclaimed that while evolution may be legitimate in principle, the theory, in fact, had not been proven and might well be entirely wrong. One gets the strong impression, moreover, that Pius was rooting pretty hard for a verdict of falsity.

Continuing directly from the last quotation, Pius advises us about the proper study of evolution:

> However, this must be done in such a way that the reasons for both opinions, that is, those favorable and those unfavorable to evolution, be weighed and judged with the necessary seriousness, moderation and measure. . . . Some however, rashly transgress this liberty of discussion, when they act as if the origin of the human body from preexisting and living matter were already completely certain and proved by the facts which have been discovered up to now and by reasoning on those facts, and as if there were nothing in the sources of divine revelation which demands the greatest moderation and caution in this question.

To summarize, Pius generally accepts the NOMA principle of nonoverlapping magisteria in permitting Catholics to entertain the hypothesis of evolution for the human body so long as they accept the divine infusion of the soul. But he then offers some (holy) fatherly advice to scientists about the status of evolution as a scientific concept: the idea is not yet proven, and you all need to be especially cautious because evolution raises many troubling issues right on the border of my magisterium. One may read this second theme in two rather different ways: either as a gratuitous incursion into a different magisterium, or as a helpful perspective from an intelligent and concerned outsider. As a man of goodwill, and in the interest of conciliation, I am content to embrace the latter reading.

In any case, this rarely quoted second claim (that evolution remains both unproven and a bit dangerous)—and not the familiar first argument for the NOMA principle (that Catholics may accept the evolution of the body so long as they embrace the creation of the soul)—defines the novelty and the interest of John Paul's recent statement.

John Paul begins by summarizing Pius's older encyclical of 1950, and particularly by reaffirming the NOMA principle—nothing new here, and no cause for extended publicity:

> In his encyclical "Humani Generis" (1950) my predecessor Pius XII had already stated that there was no opposition between evolution and the doctrine of the faith about man and his vocation.

To emphasize the power of NOMA, John Paul poses a potential problem and a sound resolution: How can we possibly reconcile science's claim for physical continuity in human evolution with Catholicism's insistence that the soul must enter at a moment of divine infusion?

> With man, then, we find ourselves in the presence of an ontological difference, an ontological leap, one could say. However, does not the posing of such ontological discontinuity run counter to that physical continuity which seems to be the main thread of research into evolution in the field of physics and chemistry? Consideration of the method used in the various branches of knowledge makes it possible to reconcile two points of view which would seem irreconcilable. The sciences of observation describe and measure the multiple manifestations of life with increasing precision and correlate them with the time line. The moment of transition to the spiritual cannot be the object of this kind of observation.

The novelty and news value of John Paul's statement lies, rather, in his profound revision of Pius's second and rarely quoted claim that evolution, while conceivable in principle and reconcilable with religion, can cite little persuasive evidence in support, and may well be false. John Paul states—and I can only say amen, and thanks for noticing—that the half century between Pius surveying the ruins of World War II and his own pontificate heralding the dawn of a new millennium has witnessed such a growth of data, and such a refinement of theory, that evolution can no longer be doubted by people of goodwill and keen intellect:

> Pius XII added . . . that this opinion [evolution] should not be adopted as though it were a certain, proven doctrine . . . Today, almost half a century after the publication of the encyclical, new knowledge has led to the recognition of the theory of evolution as more than a hypothesis.[2] It is indeed remarkable that this theory has been progressively accepted by researchers, following a series of discoveries in various fields of knowledge. The convergence, neither sought nor fabricated, of the results of work that was conducted independently is in itself a significant argument in favor of the theory.

In conclusion, Pius had grudgingly admitted evolution as a legitimate hypothesis that he regarded as only tentatively supported and potentially (as he

clearly hoped) untrué. John Paul, nearly fifty years later, reaffirms the legitimacy of evolution under the NOMA principle—no news here—but then adds
that additional data and theory have placed the factuality of evolution beyond
reasonable doubt. Sincere Christians must now accept evolution not merely as
a plausible possibility, but also as an effectively proven fact. In other words,
official Catholic opinion on evolution has moved from "say it ain't so, but we
can deal with it if we have to" (Pius's grudging view of 1950) to John Paul's
entirely welcoming "it has been proven true; we always celebrate nature's factuality, and we look forward to interesting discussions of theological implications." I happily endorse this turn of events as gospel—literally good news. I
may represent the magisterium of science, but I welcome the support of a primary leader from the other major magisterium of our complex lives. And I
recall the wisdom of King Solomon: "As cold waters to a thirsty soul, so is good
news from a far country" (Prov. 25:25).

Just as religion must bear the cross of its hard-liners, I have some scientific
colleagues, including a few in prominent enough positions to wield influence by
their writings, who view this rapprochement of the separate magisteria with dis-
may. To colleagues like me—agnostic scientists who welcome and celebrate the
rapprochement, especially the pope's latest statement—they say, "C'mon, be
honest; you know that religion is addlepated, superstitious, old-fashioned BS.
You're only making those welcoming noises because religion is so powerful, and
we need to be diplomatic in order to buy public support for science." I do not
think that many scientists hold this view, but such a position fills me with dis-
may—and I therefore end this essay with a personal statement about religion,
as a testimony to what I regard as a virtual consensus among thoughtful scientists (who support the NOMA principle as firmly as the pope does).

I am not, personally, a believer or a religious man in any sense of institutional commitment or practice. But I have great respect for religion, and the
subject has always fascinated me, beyond almost all others (with a few exceptions, like evolution and paleontology). Much of this fascination lies in the
stunning historical paradox that organized religion has fostered, throughout
Western history, both the most unspeakable horrors and the most heartrending
examples of human goodness in the face of personal danger. (The evil, I believe,
lies in an occasional confluence of religion with secular power. The Catholic
Church has sponsored its share of horrors, from Inquisitions to liquidations—
but only because this institution held great secular power during so much of

Western history. When my folks held such sway, more briefly and in Old Testament times, we committed similar atrocities with the same rationales.)

I believe, with all my heart, in a respectful, even loving, concordat between our magisteria—the NOMA concept. NOMA represents a principled position on moral and intellectual grounds, not a merely diplomatic solution. NOMA also cuts both ways. If religion can no longer dictate the nature of factual conclusions residing properly within the magisterium of science, then scientists cannot claim higher insight into moral truth from any superior knowledge of the world's empirical constitution. This mutual humility leads to important practical consequences in a world of such diverse passions.

Religion is too important for too many people to permit any dismissal or denigration of the comfort still sought by many folks from theology. I may, for example, privately suspect that papal insistence on divine infusion of the soul represents a sop to our fears, a device for maintaining a belief in human superiority within an evolutionary world offering no privileged position to any creature. But I also know that the subject of souls lies outside the magisterium of science. My world cannot prove or disprove such a notion, and the concept of souls cannot threaten or impact my domain. Moreover, while I cannot personally accept the Catholic view of souls, I surely honor the metaphorical value of such a concept both for grounding moral discussion, and for expressing what we most value about human potentiality: our decency, our care, and all the ethical and intellectual struggles that the evolution of consciousness imposed upon us.

As a moral position (and therefore not as a deduction from my knowledge of nature's factuality), I prefer the "cold bath" theory that nature can be truly "cruel" and "indifferent" in the utterly inappropriate terms of our ethical discourse—because nature does not exist for us, didn't know we were coming (we are, after all, interlopers of the latest geological moment), and doesn't give a damn about us (speaking metaphorically). I regard such a position as liberating, not depressing, because we then gain the capacity to conduct moral discourse—and nothing could be more important—in our own terms, free from the delusion that we might read moral truth passively from nature's factuality.

But I recognize that such a position frightens many people, and that a more spiritual view of nature retains broad appeal (acknowledging the factuality of evolution, but still seeking some intrinsic meaning in human terms, and from the magisterium of religion). I do appreciate, for example, the struggles of a man who wrote to the *New York Times* on November 3, 1996, to declare both his pain and his endorsement of John Paul's statement:

> Pope John Paul II's acceptance of evolution touches the doubt in my heart. The problem of pain and suffering in a world created by a God who is all love and light is hard enough to bear, even if one is a creationist. But at least a creationist can say that the original creation, coming from the hand of God, was good, harmonious, innocent and gentle. What can one say about evolution, even a spiritual theory of evolution? Pain and suffering, mindless cruelty and terror are its means of creation. Evolution's engine is the grinding of predatory teeth upon the screaming, living flesh and bones of prey. . . . If evolution be true, my faith has rougher seas to sail.

I don't agree with this man, but we could have a terrific argument. I would push the "cold bath" theory; he would (presumably) advocate the theme of inherent spiritual meaning in nature, however opaque the signal. But we would both be enlightened and filled with better understanding of these deep and ultimately unanswerable issues. Here, I believe, lies the greatest strength and necessity of NOMA, the nonoverlapping magisteria of science and religion. NOMA permits—indeed enjoins—the prospect of respectful discourse, of constant input from both magisteria toward the common goal of wisdom. If human beings can lay claim to anything special, we evolved as the only creatures that must ponder and talk. Pope John Paul II would surely point out to me that his magisterium has always recognized this uniqueness, for John's gospel begins by stating *in principio erat verbum*—in the beginning was the word.

Notes

1. Interestingly, the main thrust of these paragraphs does not address evolution in general, but lies in refuting a doctrine that Pius calls "polygenism," or the notion of human ancestry from multiple parents—for he regards such an idea as incompatible with the doctrine of original sin "which proceeds from a sin actually committed by an individual Adam and which, through generation, is passed on to all and is in everyone as his own." In this one instance, Pius may be transgressing the NOMA principle—but I cannot judge, for I do not understand the details of Catholic theology and therefore do not know how symbolically such a statement may be read. If Pius is arguing that we cannot entertain a theory about derivation of all modern humans from an ancestral population rather than through an ancestral individual (a

potential fact) because such an idea would question the doctrine of original sin (a theological construct), then I would declare him out of line for letting the magisterium of religion dictate a conclusion within the magisterium of science.

2. This passage, here correctly translated, provides a fascinating example of the subtleties and inherent ambiguities in rendering one language into another. Translation may be the most difficult of all arts, and meanings have been reversed (and wars fought) for perfectly understandable reasons.

The pope originally issued his statement in French, where this phrase read ". . . *de norsvelles connaissances conduisent á reconnaitre dans la théorie de l'évolution plus qu'une hypothése.*" *L'Osservatore Romano*, the official Vatican newspaper, translated this passage as: "new knowledge has led to the recognition of more than one hypothesis in the theory of evolution." This version (obviously, given the official Vatican source) then appeared in all English commentaries, including the original version of this essay.

I included this original translation, but I was profoundly puzzled. Why should the pope be speaking of *several* hypotheses within the framework of evolutionary theory? But I had no means to resolve my confusion, so I assumed that the pope had probably fallen under the false impression (a fairly common misconception) that, although evolution had been documented beyond reasonable doubt, natural selection had fallen under suspicion as a primary mechanism, while other alternatives had risen to prominence.

Other theologians and scientists were equally puzzled, leading to inquiries and a resolution of the problem as an error in translation (as many of us would have realized right away if we had seen the original French, or even known that the document had been issued in French). The problem lies with ambiguity in double meaning for the indefinite article in French—where *un* (feminine *une*) can mean either "a" or "one." Clearly, the pope had meant that the theory of evolution had now become strong enough to rank as "more than a hypothesis" (*plus qu'une hypothése*), but the Vatican originally read *une* as "one" and gave the almost opposite rendition: "more than *one* hypothesis." Caveat emptor.

I thank about a dozen correspondents for pointing out this error, and the Vatican's acknowledgment, to me. I am especially grateful to Boyce Rensberger, one of America's most astute journalists on evolutionary subjects, and David M. Byers, executive director of the National Conference of Catholic Bishops' Committee on Science and Human Values. Byers affirms the NOMA principle by writing to me: "Thank you for your recent article. . . . It admirably captures the relationship between science and religion that the Catholic Bishops' Committee works to promote and to realize. The text of the October 1996 papal statement from which you were working contains a mistranslation of a key phrase; the correct translation supports your thesis with even greater force."

22

YOU CAN'T HAVE IT BOTH WAYS
Irreconcilable Differences?

RICHARD DAWKINS

A cowardly flabbiness of the intellect afflicts otherwise rational people confronted with long-established religions (though, significantly, not in the face of younger traditions such as Scientology or the Moonies). S. J. Gould, commenting on the pope's attitude to evolution, is representative of a dominant strain of conciliatory thought, among believers and nonbelievers alike:

> Science and religion are not in conflict, for their teachings occupy distinctly different domains . . . I believe, with all my heart, in a respectful, even *loving* concordat [my emphasis]. . . .

Well, what are these two distinctly different domains, these "Nonoverlapping Magisteria" which should snuggle up together in a respectful and loving concordat? Gould again:

Richard Dawkins's article originally appeared as "Obscurantism to the Rescue" in the *Quarterly Review of Biology* 72 (1997), © University of Chicago Press, as one of a series of four commissioned commentaries on the Pope's Message on Evolution. Under the author's preferred title, "You Can't Have It Both Ways: Irreconcilable Differences?" it was republished in the *Skeptical Inquirer* 23, no. 4 (July/August 1999).

The net of science covers the empirical universe: what is it made of
(fact) and why does it work this way (theory). The net of religion extends over questions of moral meaning and value.

Would that it were that tidy. In a moment I'll look at what the pope
actually says about evolution, and then at other claims of his church, to see if
they really are so neatly distinct from the domain of science. First though, a
brief aside on the claim that religion has some special expertise to offer us on
moral questions. This is often blithely accepted even by the nonreligious, presumably in the course of a civilized "bending over backwards" to concede the
best point your opponent has to offer—however weak that best point may be.

The question, "What is right and what is wrong?" is a genuinely difficult
question which science certainly cannot answer. Given a moral premise or a
priori moral belief, the important and rigorous discipline of secular moral philosophy can pursue scientific or logical modes of reasoning to point up hidden
implications of such beliefs, and hidden inconsistencies between them. But the
absolute moral premises themselves must come from elsewhere, presumably
from unargued conviction. Or, it might be hoped, from religion—meaning
some combination of authority, revelation, tradition, and scripture.

Unfortunately, the hope that religion might provide a bedrock, from
which our otherwise sand-based morals can be derived, is a forlorn one. In
practice no civilized person uses scripture as ultimate authority for moral reasoning. Instead, we pick and choose the nice bits of scripture (like the Sermon
on the Mount) and blithely ignore the nasty bits (like the obligation to stone
adulteresses, execute apostates, and punish the grandchildren of offenders).
The God of the Old Testament himself, with his pitilessly vengeful jealousy, his
racism, sexism, and terrifying bloodlust, will not be adopted as a literal role
model by anybody you or I would wish to know. Yes, *of course* it is unfair to
judge the customs of an earlier era by the enlightened standards of our own.
But that is precisely my *point!* Evidently, we have some alternative source of
ultimate moral conviction which overrides scripture when it suits us.

That alternative source seems to be some kind of liberal consensus of
decency and natural justice which changes over historical time, frequently
under the influence of secular reformists. Admittedly, that doesn't sound like
bedrock. But in practice we, including the religious among us, give it higher
priority than scripture. In practice we more or less ignore scripture, quoting it
when it supports our liberal consensus, quietly forgetting it when it doesn't.

And, wherever that liberal consensus comes from, it is available to all of us, whether we are religious or not.

Similarly, great religious teachers like Jesus or Gautama Buddha may inspire us, by their good example, to adopt their personal moral convictions. But again we pick and choose among religious leaders, avoiding the bad examples of Jim Jones or Charles Manson, and we may choose good secular role models such as Jawaharlal Nehru or Nelson Mandela. Traditions too, however anciently followed, may be good or bad, and we use our secular judgment of decency and natural justice to decide which ones to follow, which to give up.

But that discussion of moral values was a digression. I now turn to my main topic of evolution, and whether the pope lives up to the ideal of keeping off the scientific grass. His Message on Evolution to the Pontifical Academy of Sciences begins with some casuistical double-talk designed to reconcile what John Paul is about to say with the previous, more equivocal pronouncements of Pius XII whose acceptance of evolution was comparatively grudging and reluctant. Then the pope comes to the harder task of reconciling scientific evidence with "revelation."

> Revelation teaches us that [man] was created in the image and likeness of God . . . if the human body takes its origin from pre-existent living matter, the spiritual soul is immediately created by God. . . . Consequently, theories of evolution which, in accordance with the philosophies inspiring them, consider the mind as emerging from the forces of living matter, or as a mere epiphenomenon of this matter, are incompatible with the truth about man. . . . With man, then, we find ourselves in the presence of an ontological difference, an ontological leap, one could say.

To do the pope credit, at this point he recognizes the essential contradiction between the two positions he is attempting to reconcile:

> However, does not the posing of such ontological discontinuity run counter to that physical continuity which seems to be the main thread of research into evolution in the field of physics and chemistry?

Never fear. As so often in the past, obscurantism comes to the rescue:

> Consideration of the method used in the various branches of knowledge makes it possible to reconcile two points of view which would

seem irreconcilable. The sciences of observation describe and measure the multiple manifestations of life with increasing precision and correlate them with the time line. The moment of transition to the spiritual cannot be the object of this kind of observation, which nevertheless can discover at the experimental level a series of very valuable signs indicating what is specific to the human being.

In plain language, there came a moment in the evolution of hominids when God intervened and injected a human soul into a previously animal lineage (When? A million years ago? Two million years ago? Between *Homo erectus* and *Homo sapiens?* Between 'archaic' *Homo sapiens* and *H. sapiens sapiens?*). The sudden injection is necessary, of course, otherwise there would be no distinction upon which to base Catholic morality, which is speciesist to the core. You can kill adult animals for meat, but abortion and euthanasia are murder because *human* life is involved.

Catholicism's "net" is not limited to moral considerations, if only because Catholic morals have scientific implications. Catholic morality demands the presence of a great gulf between *Homo sapiens* and the rest of the animal kingdom. Such a gulf is fundamentally antievolutionary. The sudden injection of an immortal soul in the time line is an antievolutionary intrusion into the domain of science.

More generally it is completely unrealistic to claim, as Gould and many others do, that religion keeps itself away from science's turf, restricting itself to morals and values. A universe with a supernatural presence would be a fundamentally and qualitatively different kind of universe from one without. The difference is, inescapably, a scientific difference. Religions make existence claims, and this means scientific claims.

The same is true of many of the major doctrines of the Roman Catholic Church. The Virgin Birth, the bodily Assumption of the Blessed Virgin Mary, the Resurrection of Jesus, the survival of our own souls after death: these are all claims of a clearly scientific nature. Either Jesus had a corporeal father or he didn't. This is not a question of "values" or "morals," it is a question of sober fact. We may not have the evidence to answer it, but it is a scientific question, nevertheless. You may be sure that, if any evidence supporting the claim were discovered, the Vatican would not be reticent in promoting it.

Either Mary's body decayed when she died, or it was physically removed from this planet to Heaven. The official Roman Catholic doctrine of Assumption, promulgated as recently as 1950, implies that Heaven has a physical location and exists in the domain of physical reality—how else could the physical body of a woman go there? I am not, here, saying that the doctrine of the Assumption of the Virgin is necessarily false (although of course I think it is). I am simply rebutting the claim that it is outside the domain of science. On the contrary, the Assumption of the Virgin is transparently a scientific theory. So is the theory that our souls survive bodily death and so are all stories of angelic visitations, Marian manifestations, and miracles of all types.

There is something dishonestly self-serving in the tactic of claiming that all religious beliefs are outside the domain of science. On the one hand miracle stories and the promise of life after death are used to impress simple people, win converts, and swell congregations. It is precisely their scientific power that gives these stories their popular appeal. But at the same time it is considered below the belt to subject the same stories to the ordinary rigors of scientific criticism: these are religious matters and therefore outside the domain of science. But you cannot have it both ways. At least, religious theorists and apologists should not be allowed to get away with having it both ways. Unfortunately all too many of us, including nonreligious people, are unaccountably ready to let them get away with it.

I suppose it is gratifying to have the pope as an ally in the struggle against fundamentalist creationism. It is certainly amusing to see the rug pulled out from under the feet of Catholic creationists such as Michael Behe. Even so, given a choice between honest-to-goodness fundamentalism on the one hand, and the obscurantist, disingenuous doublethink of the Roman Catholic Church on the other, I know which I prefer.

23
WHERE THE
TWO WORLDS TANGLE
There Is a Conflict in Metaphysics—
but Not in Ethics

Richard Feynman

There is a difficulty that the student has when he studies science, and which is, in a measure, a kind of conflict between science and religion, because it is a human difficulty that happens when you are educated two ways. Although we may argue theologically and on a high-class philosophical level that there is no conflict, it is still true that the young man who comes from a religious family gets into some argument with himself and his friends when he studies science, so there is some kind of a conflict.

The second origin of this type of conflict is associated with the facts, or, more carefully, the partial facts that he learns in the science. For example, he learns about the size of the universe. The size of the universe is very impressive, with us on a tiny particle that whirls around the Sun. That's one sun among a hundred thousand million suns in this galaxy, itself among a billion galaxies. And again, he learns about the close biological relationship of man to the animals and of one form of life to another and that man is a latecomer in a long

Richard Feynman's "Where the Two Worlds Tangle: There Is a Conflict in Metaphysics—but Not in Ethics" is an excerpt from *The Meaning of It All: Thoughts of a Citizen Scientist* (Helix Books/Addison-Wesley Books/Perseus Books, ©1998 by Michelle Feynman and Carl Feynman). The book is a newly discovered three-part lecture given at the University of Washington.

and vast, evolving drama. Can the rest be just a scaffolding for His creation? And yet again there are the atoms, of which all appears to be constructed following immutable laws. Nothing can escape it. The stars are made of the same stuff, the animals are made of the same stuff—but in some such complexity as to mysteriously appear alive.

It is a great adventure to contemplate the universe, beyond man, to contemplate what it would be like without man, as it was in a great part of his long history and as it is in a great majority of places. When this objective view is finally attained, and the mystery and majesty of matter are fully appreciated to then turn the objective eye back on man viewed as matter, to see life as part of this universal mystery of greatest depth, is to sense an experience that is very rare, and very exciting. It usually ends in laughter and a delight in the futility of trying to understand what this atom in the universe is, this thing—atoms with curiosity—that looks at itself and wonders why it wonders. Well, these scientific views end in awe and mystery, lost at the edge in uncertainty, but they appear to be so deep and so impressive that the theory that it is all arranged as a stage for God to watch man's struggle for good and evil seems inadequate.

Some will tell me that I have just described a religious experience. Very well, you may call it what you will. Then, in that language I would say that the young man's religious experience is of such a kind that he finds the religion of his church inadequate to describe, to encompass that kind of experience. The God of the church isn't big enough.

Perhaps. Everyone has different opinions.

Suppose, however, our student does come to the view that individual prayer is not heard. I am not trying to disprove the existence of God. I am only trying to give you some understanding of the origin of the difficulties that people have who are educated from two different points of view. It is not possible to disprove the existence of God, as far as I know. But it is true that it is difficult to take two different points of view that come from different directions. So let us suppose that this particular student is particularly difficult and does come to the conclusion that individual prayer is not heard. Then what happens? Then the doubting machinery, his doubts, are turned on ethical problems. Because, as he was educated, his religious views had it that the ethical and moral values were the word of God. Now if God maybe isn't there, maybe the ethical and moral values are wrong. And what is very interesting is that they have survived almost intact. There may have been a period when a few of the moral views and the ethical positions of his religion seemed wrong, when he had to think about them, but many of them he returned to.

But my atheistic scientific colleagues, which does not include all scientists—I cannot tell by their behavior, because of course I am on the same side, that they are particularly different from the religious ones—it seems that their moral feelings and the understandings of other people and their humanity and so on apply to the believers as well as the disbelievers. It seems to me that there is a kind of independence between the ethical and moral views and the theory of the machinery of the universe.

Science, indeed, makes an impact on many ideas associated with religion, but I do not believe it affects, in any very strong way, the moral conduct and ethical views. Religion has many aspects. It answers all kinds of questions. I would, however, like to emphasize three aspects.

The first is that it tells what things are and where they came from and what man is and what God is and what properties God has and so on. I'd like, for the purposes of this discussion, to call those the metaphysical aspects of religion.

And then it says how to behave. I don't mean in the terms of ceremonies or rituals or things like that, but I mean how to behave in general, in a moral way. This we could call the ethical aspect of religion.

And finally, people are weak. It takes more than the right conscience to produce right behavior. And even though you may feel you know what to do, you all know that you don't do the way you would like yourself to do. And one of the powerful aspects of religion is its inspirational aspects. Religion gives inspiration to act well. Not only that, it gives inspiration to the arts and to many other activities of human beings.

Now these three aspects of religion are very closely interconnected, in the religions' view. First of all, it usually goes something like this: that the moral values are the word of God. Being the word of God connects the ethical and metaphysical aspects of religion. And finally, that also inspires the inspiration, because if you are working for God and obeying God's will, you are in some way connected to the universe, your actions have a meaning in the greater world, and that is an inspiring aspect. So these three aspects are very well integrated and interconnected. The difficulty is that science occasionally conflicts with the first two categories, that is with the ethical and with the metaphysical aspects of religion.

There was a big struggle when it was discovered that the Earth rotates on its axis and goes around the Sun. It was not supposed to be the case according to the religion of the time. There was a terrible argument and the outcome was, in that case, that religion retreated from the position that the Earth stood at the center of the universe. But at the end of the retreat there was no change in the moral viewpoint of the religion. There was another tremendous argument when it was found likely that man descended from the animals. Most religions have retreated once again from the metaphysical position that it wasn't true. The result is no particular change in the moral view. You see that the Earth moves around the Sun, yes, then does that tell us whether it is or is not good to turn the other cheek? It is this conflict associated with these metaphysical aspects that is doubly difficult because the facts conflict. Not only the facts, but the spirits conflict. Not only are there difficulties about whether the Sun does or doesn't rotate around the Earth, but the spirit or attitude toward the facts is also different in religion from what it is in science. The uncertainty that is necessary in order to appreciate nature is not easily correlated with the feeling of certainty in faith, which is usually associated with deep religious belief. I do not believe that the scientist can have that same certainty of faith that very deeply religious people have. Perhaps they can. I don't know. I think that it is difficult. But anyhow it seems that the metaphysical aspects of religion have nothing to do with the ethical values, that the moral values seem somehow to be outside the scientific realm.

V

The Scientific Investigation of
Paranatural Claims

24
EXAMINING CLAIMS OF
THE "PARANATURAL"
Life After Death

PAUL KURTZ

What Is the "Paranatural"?

Can science investigate and unravel claims made on behalf of theistic religion? My answer is in the affirmative; however, it will be most effective if it interprets such events in *paranatural* terms; that is, if it translates purely metaphysical or speculative transcendental claims into testable hypotheses.

What do I mean by the term *paranatural*? Science presupposes naturalism; that is, it seeks to develop causal explanations of natural phenomena, and it tests its hypotheses and theories by reference to the principles of logic, empirical observation, experimental prediction, and confirmation.

This is in contrast with supernatural explanations, which claim to deal with an order of existence beyond the visible or observable universe, and attribute events to occult causes. Supernaturalism postulates divine powers intervening miraculously in natural causal sequences. Thus it is alleged that the

Paul Kurtz's "Examining Claims of the 'Paranatural': Life After Death" was an invited paper delivered at the annual meeting of the American Physical Society, held in Minneapolis on March 22, 2000. A modified version later appeared in the *Skeptical Inquirer* 24, no. 6 (November/December 2000).

natural and material universe needs to be supplemented by a supernatural reality, which transcends human understanding and can only be approached by mysticism and faith. The domain of faith, it is said, supplements the domain of reason.

There are at least two classes of events that stand between the natural and supernatural realms and enable us in some sense to deal with the occult. These refer to (1) paranormal and (2) paranatural phenomena. The term *paranormal* was used in the past century by parapsychologists (such as J. B. Rhine and Samuel Soal) to refer to a class of anomalous events that its proponents claimed were inexplicable in terms of normal materialistic sciences. "Para" meant "besides, alongside of, or beyond" naturalistic psychology. Nonetheless, these parapsychologists maintained that it was possible to describe and perhaps interpret these events experimentally, and they did so by referring to a range of psi phenomena, which referred to ESP, telepathy, clairvoyance, precognition, and psychokinesis.

There is another range of events, which I have labeled as paranatural, that deal with still other dimensions of reality: classical mystical or supernatural claims that allegedly intrude into our universe from without. I am here referring primarily to a theistic order of reality and to phenomena including discarnate souls, intelligent design, and "creation science." Visitations from extraterrestrials beyond this world may be considered to be both paranormal and paranatural. Included under this rubric of "paranatural" are some classical religious phenomena, such as weeping statues, stigmata, exorcism and possession, faith healing, the Shroud of Turin, past-life regressions used as evidence for reincarnation, historical revelations by prophets who carry messages from On High, and other so-called religious miracles. All of these have an empirical component and are not completely transcendental, and hence they are capable of some experimental testing and historical reconstruction of their claims. Although these anomalous events are beyond nature, in one sense, proponents of them seek to offer some kind of empirical evidence to support their belief that there are nonnatural, nonmaterial, or spiritual processes at work in the universe.

I disagree with the claims of the defenders of the *para*: I do not think that either the paranormal or paranatural exist outside of nature or that they constitute dimensions of reality that undermine naturalism. *Para* is a substitute for our ignorance at any one time in history (as is the term *miracle*, which is interjected when we do not understand the causes of phenomena). Indeed, as we expand the frontiers of knowledge, phenomena considered *para* can, I submit, be given naturalistic or normal explanations, and this range of phenomena can either be interpreted by the existing body of explanatory scientific principles or by the introduction of new ones.

Life After Death

I wish to illustrate this by dealing with the intriguing question: What is the evidence for life after death? Can we communicate with the dead? That is, Are we able to be in touch with people who have died? Do they have some form of existence, perhaps as "discarnate spirits" or "disembodied souls"? This is an age-old question that is related to faith in immortality and a very deep hunger for it. Although it has been interpreted as "paranormal," it may more appropriately be considered to be "paranatural" because of its religious significance. Indeed, for the major supernatural religions of the world—Christianity, Judaism, and Islam—belief in an afterlife and the promise of heaven are central.

At present there is intense popular interest in these questions in the United States. It is stimulated by the mass media, at least as measured by the number of popular books, magazine articles, movies, and television and radio programs devoted to the theme. The films *The Sixth Sense* (with Bruce Willis and Haley Joel Osment) and *Poltergeist* are examples of the prevailing interest, as are the best-selling books by James Van Praagh (*Talking to Heaven*, 1997; *Reaching to Heaven*, 1999), John Edward (*One Last Time*, 1998), Sylvia Browne (*The Other Side and Back*, 1999), and Rosemary Altea (*You Own the Power*, 1999). Dan Rather on CBS, the Fox TV network, *Larry King Live*, and some talk-show hosts have devoted many uncritical programs to these claims. For example, the HBO TV network did a special recently, "Life Afterlife," purporting to present the scientific examination of survival. It interviewed dozens of people, all of whom claim to have communicated with the dead, and several parapsychologists, all arguing the case for survival. There are all too few objective programs examining such questions; most favor a spiritual-paranormal interpretation.

As a result of a massive media onslaught, polls in the last decade place the United States as number one in belief in life after death in the democratic world, and higher than virtually all European countries. Two cross-national surveys conducted for the International Social Survey Program in 1991 and 1993[1] indicate that the United States ranked highest, along with Ireland and the Philippines, for those who believe in heaven (63.17 percent of the population), for those who believe in hell (49.6 percent), and for those who believe in life after death (55 percent). The United States was lowest of twenty-one nations on knowledge of human evolution (44.2 percent), lower than Poland and Russia. Recent polls have shown the level of credulity growing in the past decade. A poll conducted for *Newsweek* magazine by the Princeton Survey Research Associates, based on a sample of 752 adults interviewed, indicated that 84 percent of Americans said that God performed miracles and 77 percent said saints or God can cure people otherwise medically incurable.[2] Paradoxically, the United States is allegedly the most advanced scientific-technological society in the world.

A Brief History of Life after Death Claims

What do scientists have to say about life after death? Science has been investigating our ability to communicate with the dead for at least 150 years and it has attempted to discover empirical evidence in support of the claim. It began to do so with the emergence of spiritualism in the nineteenth century with the Fox sisters (Margaret and Kate), two young girls in Hydesville, New York (outside of Rochester). In 1848 they first claimed that they could receive messages from "the spirit world beyond." In their presence, there were strange rappings; people would receive answers to their questions spelled out by the number of taps (Kurtz 1985). The basic premise was that human personality survived death and could communicate with specially endowed mediums. In the late nineteenth century and early twentieth century spiritualism swept the United States, England, and Europe. Thousands of mediums soon appeared, all seemingly capable of communicating with the dead. The most popular method of investigation was to try to communicate in a specially darkened séance room, wherein the discarnate entity would make its presence known by physical or verbal manifestations: table tipping, levitation of objects, ectoplasmic emissions, teleportation, materializations, automatic writings, and the like.

A committee of medical doctors at the University of Buffalo tested the Fox sisters in 1851 and attributed their raps to the cracking of their toe knuckles or kneecaps against a wooden floor or bedstead. The physicians did a controlled experiment by placing the girls' feet on pillows, and nothing happened. The great physicist Michael Farraday investigated table tipping (1852) and found that it was due to pressure exerted by the fingers of the sitters (whether voluntarily or involuntarily). Sir Walter Crookes investigated the most colorful mediums of the day, D. D. Home (1871) and Florence Cook (1873), and thought that they had special abilities of mediumship—though critics believe that he was duped by both (Hall 1962, 1984).

The Society for Psychical Research was founded in 1882 in Great Britain by Henry Sidgwick, Richard Hodgson, F. W. H. Myers, Edmund Gurney, and others to investigate survival of life after death, among other questions. The American branch of the Society for Psychical Research was founded in 1885 by William James at Harvard University. These researchers examined reports of apparitions and ghostly hauntings. It was difficult to corroborate these subjective eyewitness accounts and so these investigations focused on physical manifestations. There were numerous photographs of ghosts—which it was soon discovered could easily be doctored. Many famous mediums such as Eusapia Palladino (in Italy) and Leonora Piper (in Boston) were tested under controlled conditions in an effort to determine whether they possessed extraordinary powers.

Palladino was especially elusive, and the scientific community was split as to whether she was fraudulent. The Feilding Report was an account of sittings done in Naples (1909) by a team of scientists who thought she was genuine. Palladino was also tested in the United States at Harvard by Hugo Muensterberg (1909) and at Columbia University (1910) by a team of scientists; and in both cases the physical levitation of the table behind her and the feeling of being pinched by her spirit control (called John King) was found to be caused by her adroit ability to stretch her leg in contortions and to pinch sitters with her toes, or to raise a small table behind her so as to appear to levitate. This was detected by having a man dressed in black crawl under the table and see her at work. A subsequent Feilding report (1911) also found that she had cheated (Kurtz 1985).

Late in his career the famous magician Houdini (1874–1926) exposed several bogus mediums. By the 1920s the spiritualist movement was thoroughly discredited, because when the controls were tightened, the effect disappeared; skeptics insisted that if a person claims to be in contact with a spiritual entity, there must be some independent physical corroboration by impartial observers (Houdini 1924, 1981).

In the 1930s the survival question in science was laid aside. J. B. Rhine and others focused instead on psi phenomena, again with controversial results, because scientists demanded replicable experiments by neutral observers, which were difficult to come by (Hansel 1980). In any case, whether or not psi existed was independent of the survival question.

Spiritualism Returns

In recent decades interest in the survival question has reappeared. This is rather surprising to skeptical investigators. No doubt this revival of interest is due in part to the growth of religiosity and spirituality on the broader cultural scene, but it is also due to the sensationalism of the mass media. I can only briefly outline some of the claims that have been made and the kinds of research that have been done. Most of this work is highly questionable, for the standards of rigorous methodological inquiry so essential to science seem to have declined drastically from what occurred in the early part of the last century.

(1) **Channeling to the other side.** Surprisingly, a new class of mediums, now called "channelers," have emerged (such as James Van Praagh, John Edward, Sylvia Browne, and Rosemary Altea previously cited) who claim to be able to be put themselves into immediate contact with a dead relative or friend and to convey a message back from them. Thus, what we have are subjective reports based on the word of the channeler that he or she is in touch with the departed spirit. There are two ways that this is done. First, there are "hot readings," when the channeler may know something by previous research about the person being read. A good case of this is Arthur Ford, who did a reading of Bishop James Pike and claimed he was in contact with his son who had committed suicide. It was discovered after Ford's death that he had done an extensive background investigation of Pike's son before the reading. The most common method used, however, is the skillful use of "cold readings" by the channeler. The public here is taken in by flim-flammery, and there is all too little effort to critically examine the claims made.

There has been a massive shift in the methodology used. If in previous decades scientists demanded some corroborative and/or physical manifestation of mediumship, today all rigorous standards of evidence and verification seem to have been abandoned. Psychologist Ray Hyman has shown how a psychic gives a general cold reading: if he throws out messages from the spirit world to an audience someone will usually emerge to whom it fits (Hyman 1977). Thus, he may ask, "Does anyone know a Mary or a William?" And most likely a person will step forth who does, and then the reading proceeds, on a hit-and-miss basis. The skillful channeler simply has to have one or two lucky hits to mystify the audience.

(2) **Apparitions and other sightings.** Similar considerations apply to the epidemic of eyewitness testimonials that people have been reporting of ghostly apparitions, angels, and other ethereal entities. Such stories are pervasive in contemporary culture, since a tale once uttered may spread rapidly throughout the population; this is facilitated by the mass media and becomes contagious. If someone claims to see ghosts or angels, other people, perhaps millions, may likewise begin to encounter them.

What is so curious is that people who see ghosts usually see them clothed. It is one thing to say that a discarnate soul has survived, but that his or her clothing and other physical objects have survived is both amusing and contrary to the laws of physics!

The most succinct explanation that we have for this phenomenon is that it is in the eye of the beholder, satisfying some deep-felt need, a transcendental temptation or will-to-believe. The demand for independent objective verification seems to be ignored. It is puzzling why so many people will accept uncorroborated subjective reports, particularly when we find them unreliable. The death of a loved one can cause untold psychological trauma, and there are powerful motives, psychological and indeed sociological, for believing in their survival. Thus there are naturalistic psychological and sociological explanations that better account for the prevalence of such phenomenological givens, without the need to postulate discarnate beings or our ability to communicate with them.

Let me briefly outline two other areas of survival research, which at least claim to be more carefully designed.

(3) **Death-bed visions.** Osis and Haraldsson (1974, 1977) sent out questionnaires to doctors and nurses to ask them to describe the verbal accounts of death-bed visions of people in their last moments of dying. The question is whether these persons were able to communicate with departed friends or relatives at the last moment or were merely hallucinating, as skeptics suggest they were. In any case, virtually all of this data is secondhand, and is influenced by cultural expectations that when we die we will meet people on the other side.

(4) **The phenomenology of near-death experience.** This is a very popular area of research today, widely touted as evidence for communication, and based on firsthand testimony. Much research has gone into this intriguing area by Raymond Moody (1975, 1977), Elizabeth Kübler-Ross (1981), Kenneth Ring (1980, 1984, 1998), Michael Sabom (1982), and Melvin Morse (1990),

among others. These extended phenomenological reports claim to give us evidence from the other side from people who were dying and resuscitated. There is an out-of-body experience, a vision of a tunnel, a bright light, a recall of one's life, and perhaps a meeting with beings on the other side.

Critics claim that the descriptive collage offered is of the dying process, and that in no case do we have reports of persons who have died (i.e., experienced brain death) and communicated with those on the other side. There is a variety of alternative naturalistic explanations. Skeptics maintain we are most likely dealing with psychological phenomena, where the person facing death has either hallucinations, has reached a state of depersonalization, and/or there are changes in brain chemistry and the nervous system (Blackmore 1993). Some have postulated that the discarnate entities or divine beings encountered on the other side are colored by the sociocultural context (Kellehear 1996), though proponents maintain that in spite of this there is a common core of similarities. Some have said that falls or accidents where a person thinks he is about to die, but survives, can cause analogous out-of-body experiences and panoramic reviews (Noyes and Klette 1972, 1977). Not everyone who is dying reports near-death experiences; many people who are not dying report having them. Sleep paralysis and hypnopompic and hypnagogic dream states are factors in common out-of-body experiences. Ronald Siegel (1981) maintains that similar NDEs can be induced by hallucinogens. Karl Jansen (1996) has presented evidence that they can be stimulated by the dissociative drug ketamine. Various conditions can precipitate an NDE, such as low blood sugar, oxygen deprivation, reduced blood flow, temporal-lobe epilepsy, etc., and can lead to an altered state of consciousness. For skeptical inquirers, in no case can we say that the person has died and returns; what we are dealing with is the process or belief that one is dying.

Analytic philosophers have pointed out additional serious conceptual difficulties in the hypothesis that nonphysical beings are communicating with us—there is a sharp mind/body dualism here. Perhaps the real question is not whether there is sufficient evidence for "x," but the meaning of "x"; and whether we can communicate with "disembodied entities" who have a level of consciousness without sensory organs or a brain. Some have claimed that the communication is "telepathic," but the experimental evidence for telepathy is itself questionable.

Conclusion

After a quarter of a century in this field of research, I find that eyewitness testimony is notoriously unreliable, and that unless carefully controlled studies and standards are applied, people can deceive themselves and others into believing that almost anything is true and real—from past-life regression and extraterrestrial abductions to satanic infestations and near-death experiences.

What should be the posture of the scientific investigator about paranatural survival claims? Clearly, we need an open mind, and we should not reject a priori any such claim; if claims are responsibly framed they should be carefully evaluated. After a century and a half of scientific research, what are we to conclude? I submit that there is insufficient reliable or objective evidence that some individuals are able to reach another plane of existence beyond this world and/or communicate with the dead. As far as we know, the death of the body entails the death of psychological functions, consciousness, and/or the personality; and there is no reason to believe that ghosts hover and haunt and/or can communicate with us.

I realize that this flies in the face of what the preponderance of religionists wish to believe, but science should deal as best it can with what is the case, not with what we would like it to be. Unfortunately, scientific objectivity today has an uphill battle in this area in the face of media sensationalism and the enormous public fascination with paranormal and paranatural claims.

Notes

1. Currently based at the National Opinion Research Center, University of Chicago.
2. Religion News Service, April 13, 2000.

References

Altea, Rosemary. 1999. *You Own the Power: Stories and Exercises.* New York: William Morrow.

Blackmore, Susan J. 1993. *Dying to Live.* Amherst, N.Y.: Prometheus Books.

Browne, Sylvia, with Lindsay Harrison. 1999. *The Other Side and Back: A Psychic's Guide to Our World and Beyond.* New York: Dutton.

Edward, John. 1998. *One Last Time: A Psychic Medium Speaks to Those We Have Loved and Lost.* New York: Berkeley Books.

Feilding, E., W. W. Baggally, and H. Carrington. 1909. "Report on a Series of Sittings with Eusapia Palladino." *Proceedings of the Society for Psychical Research* 23: 306–569.

Feilding, E., and W. Marriott. 1911. "Report on a Further Series of Sittings with Eusapia Palladino at Naples." *Proceedings of the Society for Psychical Research* 25.

Hall, Trevor H. 1962. *The Spiritualists: The Story of Florence Cook and William Crookes.* London: Duckworth.

———. 1984. *The Enigma of Daniel Home: Medium or Fraud?* Amherst, N.Y.: Prometheus Books.

Hansel, C. E. M. 1980. *ESP and Parapsychology: A Critical Evalutaion.* Amherst, N.Y.: Prometheus Books.

Houdini, Harry. 1924. *A Magician Among the Spirits.* New York: Harper.

———. 1981. *Miracle Mongers and Their Methods: A Complete Exposé.* Amherst, N.Y.: Prometheus Books.

Hyman, Ray. 1977. "Cold Reading: How to Convince Strangers that You Know All about Them." *Zetetic (Skeptical Inquirer)* 1, no. 2.

Jansen, K. L. R. 1996. "Using Ketamine to Induce the Near-Death Experience: Mechanism of Action and Therapeutic Potential." *Yearbook of Ethnomedicine and the Study of Consciousness,* no. 4.

Kellehear, Allan. 1996. *Experiences Near Death: Beyond Medicine and Religion.* New York: Oxford University Press.

Kübler-Ross, Elisabeth. 1981. *Living with Death and Dying.* New York: Macmillan.

Kurtz, Paul, ed. 1985. *A Skeptic's Handbook of Parapsychology.* Amherst, N.Y.: Prometheus Books.

Moody, Raymond A., Jr. 1975. *Life After Life: The Investigation of a Phenomenon—Survival of Bodily Death.* Covington, Calif.: Mockingbird Books.

———. 1977. *Reflections on Life After Life.* New York: Bantam Books.

———. 1999. *The Last Laugh: A New Philosophy of Near-Death Experiences, Apparitions, and the Paranormal.* Charlottesville, Va.: Hampton Road Publishing Co.

Morse, Melvin, and Paul Perry. 1990. *Closer to the Light: Learning from the Near-Death Experiences of Children.* New York: Ballantine Books.

Neher, Andrew. 1981. *The Psychology of Transcendence.* Englewood Cliffs, N.J.: Prentice-Hall.

Noyes, Russell, Jr., and Roy Klette. 1972. "The Experience of Dying from Falls." *Omega* 3: 45–52.

———. 1977. "Depersonalization in Response to Life-threatening Danger." *Comparative Psychology* 18: 375–84.

———. 1977. "Panoramic Memory: A Response to the Threat of Death." *Omega,* 8.

Osis, Karles, and Eilendur Haraldsson. 1974. "Survey of Death Visions in India." In

Research in Parapsychology 1973, edited by W. G. Roll et al. Metuchen, N.J.: Scarecrow Press.

———. 1977. "Deathbed Observations by Physicians and Nurses: A Cross-cultural Survey." *Journal of the American Society for Psychical Research* 71: 237–59.

Ring, Kenneth. 1980. *Life at Death: A Scientific Investigation of the Near-Death Experience.* New York: Coward, McCann and Geoghegan.

———. 1984. *Heading toward Omega: In Search of Near-Death Experience.* New York: William Morrow.

———. 1998. *Lessons from the Light: What We Can Learn from the Near-Death Experience.* Reading, Mass.: Perseus Books.

Sabom, Michael B. 1982. *Recollections of Death: A Medical Investigation.* New York: Harper & Row.

Siegel, Ronald. 1981. "Life After Death." In *Science and the Paranormal: Probing the Existence of the Paranormal,* edited by G. O. Abell and B. Singer. New York: Scribners.

Van Praagh, James. 1997. *Talking to Heaven: A Medium's Message of Life after Death.* New York: Dutton.

———. 1999. *Reaching to Heaven: A Spiritual Journey through Life and Death.* New York: Dutton.

25
AFTER-DEATH
COMMUNICATION STUDIES

RICHARD WISEMAN AND CIARÁN O'KEEFFE

Schwartz, Russek, Nelson, and Barentsen (2001) recently reported two studies in which mediums appeared to be able to produce accurate information about the deceased under conditions that the authors believed "eliminate the factors of fraud, error, and statistical coincidence." Their studies were widely reported in the media as scientific proof of life after death (e.g., Matthews 2001; Chapman 2001). This paper describes some of the methodological problems associated with the Schwartz et al. studies and outlines how these problems can be overcome in future research.

Schwartz et al.'s first experiment was funded and filmed by a major U.S. television network (Home Box Office—HBO) making a documentary about the survival of bodily death. The study involved two participants (referred to as "sitters") and five well-known mediums. The first sitter was a forty-six-year-old woman who had experienced the death of over six people in the last ten years. Schwartz et al. stated that this sitter was recommended to them by a

Richard Wiseman and Ciarán O'Keefe's "After-Death Communication Studies" originally appeared as "Accuracy and Replicability of Anomalous After-Death Communication across Highly Skilled Mediums: A Critique," *Paranormal Review* 19 (July 2001), and as "A Critique of Schwartz et al.'s After-Death Communication Studies" in the *Skeptical Inquirer* 25, no. 6 (November/December 2001).

well-known researcher in ADCs (After Death Communication) who "knew of the sitter's case through her research involving spontaneous ADCs." The second sitter was a fifty-four-year-old woman who had also experienced the death of at least six people in the last ten years.

During the experiment, the sitter and medium sat on either side of a large opaque screen. The medium was allowed to "conduct the reading in his or her own way, with the restriction that they could ask only questions requiring a yes or no answer." Each sitter was asked to listen to the reading and respond to the medium's questions by saying the word "yes" or "no" out loud. The first sitter was given a reading by all five mediums; the second sitter received readings from only two of the mediums.

A few months after the experiment, both sitters were asked to assign a number between -3 (definitely an error) to +3 (definitely correct) to each of the statements made by the mediums. The sitters placed 83 percent and 77 percent of the statements into the +3 category. Schwartz et al. also reported their attempt to discover whether "intelligent and motivated persons" could guess the type of information presented by the mediums by chance alone. The investigators selected seventy statements from the readings given to the first sitter and turned them into questions. For example, if the medium had said "your father loved dancing," the question became "Who loved to dance?" Sixty-eight undergraduates were shown these questions, along with a photograph of the sitter, and asked to guess the answer. Schwartz et al. reported that the average number of items guessed correctly was just 36 percent, and argue that the high level of accuracy obtained by the mediums could not be due to chance guessing.

The first sitter was then invited back to the laboratory to take part in a second experiment. In this experiment she received readings from two of the mediums who also participated in the first study. Rather than being separated by an opaque screen, the sitter sat six feet behind the medium. In the first part of these two readings the sitter was instructed to remain completely silent. In the second part she was asked to answer "yes" or "no" to each of the medium's questions. After reviewing the readings, the sitter rated 82 percent of the mediums' statements as being "definitely correct."

The Schwartz et al. studies suffered from severe methodological problems, namely: (1) the potential for judging bias, (2) the use of an inappropriate control group, and (3) inadequate safeguards against sensory leakage. Each of these problems will be discussed in turn.

Judging Bias

During a mediumistic reading the medium usually produces a large number of statements and the sitter has to decide whether these statements accurately describe his/her deceased friends or relatives. It is widely recognized that the sitter's endorsement of such statements cannot be taken as evidence of mediumistic ability, as seemingly accurate readings can be created by a set of psychological stratagems collectively referred to as "cold reading" (Hyman 1977; Rowland 1998). It is therefore vital that any investigation into the possible existence of mediumistic ability controls for the potential effect of these stratagems. Unfortunately, the Schwartz et al. study did not contain such controls, and thus it is possible that the seemingly impressive results could have been due to cold reading.

Schwartz et al. reproduced a small part of one reading in their paper, and this transcript can be used to illustrate how cold reading could account for the outcome of the studies. In the first line of the transcript the medium said, "Now, I don't know if they [the spirits] mean this by age or by generation, but they talk about the younger male that has passed. Does that make sense to you?" The sitter answered "yes." The medium's statement is ambiguous and open to several different interpretations. When the medium mentioned the word "younger" he/she could be talking about a young child, a young man, or even someone who died young (e.g., in their forties). The sitters may be motivated to interpret such statements in such a way as to maximize the degree of correspondence with their deceased friends and relatives if, for example, they had a strong belief in the afterlife, a need to believe that loved ones have survived bodily death, or were eager to please the mediums, investigators, and the HBO film crew.

In addition, the sitters may have endorsed the readings because some statements caused them to selectively remember certain events in their lives. As a hypothetical example, let us imagine that the medium had said, "Your son was an extrovert." This statement may have caused the sitter to selectively recall certain life events (i.e., the times that her son went to parties and was very outgoing), forget other events (e.g., the times that he sat alone and didn't want to be with others), and thus assign a spuriously high accuracy rating to the statement.

Biased interpretation of ambiguous statements and selective remembering can lead to sitters endorsing contradictory statements during a reading.

Interestingly, the short transcript reproduced by Schwartz et al. contains an example of exactly this happening:

Medium: . . . your dad speaks about the loss of child. That makes sense?

Sitter: Yes.

Medium: Twice? 'Cause your father says twice.

Sitter: Yes.

Medium: Wait a minute, now he says thrice. He's saying three times. Does that make sense?

Sitter: That's correct.

Some of the statements made by the mediums may also have been true of a great many people and thus had a high likelihood of being endorsed by the sitters. For example, in the transcript the medium stated that one of the spirits was a family member, and elsewhere Schwartz et al. stated that the mediums referred to "a little dog playing ball." It is highly probable that many sitters would have endorsed both of these statements. Research has also revealed that many statements that do not appear especially general can also be true of a surprisingly large number of people. Blackmore (1994) carried out a large-scale survey in which more than 6,000 people were asked to state whether quite specific statements were true of them. More than one third of people endorsed the statement, "I have a scar on my left knee" and more than a quarter answered yes to the statement "Someone in my family is called Jack." In short, the mediums in the Schwartz et al. study may have been accurate, in part, because they simply produced statements that would have been endorsed by many sitters.

Other factors may also increase the likelihood of the sitter endorsing the mediums' statements. Clearly, the more deceased people known to the sitter, the greater chance they will have of being able to find a match for the medium's comments. Both sitters knew a relatively large number of deceased people. Both of them had experienced the death of six loved ones in the last ten years, and the first sitter reported that she believed that the mediums had contacted an additional nine of her deceased friends and relatives. Thus, the sitters' high levels of endorsement may have been due, in part, to them having a large number of deceased friends and relatives.

Control Group Biases

Schwartz et al. attempted to discover whether the seemingly high accuracy rate obtained by the mediums could have been the result of chance guesswork. However, the method developed by the investigators was inappropriate and fails to address the concerns outlined above. They selected seventy statements from the readings given to the first sitter in the first experiment and turned them into questions. For example, if the medium had said, "Your son is very good with his hands," the question became, "Who was very good with his hands?" These questions were presented to a group of undergraduates, who were asked to guess the answers. Schwartz et al. reported that the average number of items guessed correctly was just 36 percent. However, it is extremely problematic to draw any conclusions from this result due to the huge differences in the tasks given to the mediums and control group. For example, when the medium said, "Your son was very good with his hands," the sitter has to decide whether this statement matches the information that she knew about her deceased son. However, as noted above, this matching process may be biased by several factors, including her selective remembering and the biased interpretation of ambiguous statements. For example, the sitter may think back to the times that her son built model airplanes, endorse the statement, and the medium would receive a "hit." However, the control group were presented with a completely different task. They were presented with the question "Who was good with his hands?" and would only receive a "hit" if they guessed that the answer was the sitter's son. They therefore had a significantly reduced likelihood of obtaining a hit than the mediums.

Conceptually, this is equivalent to testing archery skills by having someone fire an arrow, drawing a target around wherever it lands and calling it a bull's-eye, and then testing a "control" group of other archers by asking them to hit the same bull's-eye. Clearly, the control group would not perform as well as the first archer, but the difference in performance would reflect the fact that they were presented with very different tasks, rather than a difference in their archery skills.

Psychical researchers have developed various methods to overcome the problems associated with "cold reading" when investigating claims of mediumistic ability (see Schouten 1994 for an overview). Most of these methods involve the concept of "blind judging." In a typical experiment, a small number of sitters receive a reading from a medium. The sitters are then asked to evaluate both his or her own reading (often referred to as the "target" reading) and the readings made for other sitters (referred to as "decoy" readings). If the medium is accurate then the ratings assigned to the target readings will be significantly greater than those assigned to the decoy readings. However, it is absolutely vital that the readings are judged "blind"—the sitters should be unaware of whether they are evaluating a "target" or "decoy" reading. This simple safeguard helps overcome all of the problems outlined above. Let us suppose that the medium is not in contact with the spirit world, but instead tends to use cold reading to produce seemingly accurate statements. These techniques will cause the sitters to endorse both the target and decoy readings, and thus produce no evidence for mediumistic ability. If, however, the medium is actually able to communicate with the spirits, the sitters should assign a higher rating to their "target" reading than the "decoy" readings, thus providing evidence of mediumistic ability.

It is hoped that future tests of mediumistic ability will employ the type of blind judging methods that have been developed, and frequently employed, in past tests of mediumistic ability.

However, blind judging is only one of several methodological safeguards that should be employed when testing mediumistic ability. Well-controlled tests should also obviously prevent the medium from being able to receive information about a sitter through any normal channels of communication. Unfortunately, the measures taken by Schwartz et al. to guard against various forms of potential sensory leakage appear insufficient.

Sensory Leakage

Throughout all of the readings in the first experiment, and the latter part of the readings in the second experiment, the sitter was allowed to answer "yes" or "no" to the medium's questions. These answers would have provided the mediums with two types of information that may have helped them produce more accurate statements in the remainder of the reading. First, it is very likely that the sitter's voice would have given away clues about her gender, age, and socioeconomic group. This information could cause the mediums to produce statements that have a greater likelihood of being endorsed by the sitter. For example, an older sitter is more likely to have experienced the death of their parents than a younger sitter, and certain life events are gender-specific (e.g., being pregnant, having a miscarriage, etc.). Second, the sitters' answers may have also given away other useful clues to the mediums. For example, let us imagine that the medium stated, "I am getting the impression of someone male, is that correct?" If the sitter has recently lost someone very close to her, such as a father or son, then she might answer a tearful "yes." If, however, the deceased male was an uncle that sitter didn't really know very well, then her "yes" might be far less emotional. Again, a skilled medium might be able to unconsciously use this information to produce accurate statements later in the reading. Any well-controlled test of mediumistic ability should not allow for the sitter to provide verbal feedback to the medium during the reading.

In the first part of the readings in the second experiment, the sitter was asked not to answer yes or no to any of the medium's statements. However, the experimental set-up still employed insufficient safeguards against potential sensory leakage. The medium sat facing a video camera and the sitter sat six feet behind the medium without any form of screen separating the two of them. As such, the sitter may have emitted various types of sensory signals, such as cues from her movement, breathing, odor, etc. Parapsychologists have developed elaborate procedures for eliminating potential sensory leakage between participants (e.g., Milton and Wiseman 1997). These safeguards frequently involve placing participants in separate rooms, and often the use of specially constructed sound-attenuated cubicles. Schwartz et al. appeared to have ignored these guidelines and instead allowed the sitter to interact with the medium, and/or simply seated them behind one another in the same room. Neither of these measures represent sufficient safeguards against the potential for sensory leakage.

The investigators also failed to rule out the potential for sensory leakage between the experimenters and mediums. The second sitter in the first experiment is described as being "personally known" to two of the experimenters (Schwartz and Russek). The report also described how, during the experiment, the mediums were allowed to chat with Russek in a courtyard behind the laboratory. Research into the possible existence of mediumistic ability should not allow anyone who knows the sitter to come into contact with the medium. Schwartz allowed such contact, with their only safeguard being that the mediums and Russek were not allowed to talk about matters related to the session. However, a large body of research has shown that there are many ways in which information can be unwittingly communicated, via both verbal and nonverbal means (e.g., Rosenthal and Rubin 1978). As such, the safeguards employed by

Schwartz et al. against possible sensory leakage between experimenter and mediums were insufficient.

In short, the Schwartz et al. study did not employ blind judging, employed an inappropriate control group, and had insufficient safeguards against sensory leakage. As such, it is impossible to know the degree to which their findings represent evidence for mediumistic ability. Psychical researchers have worked hard to develop robust methods for testing mediums since the 1930s (see Schouten 1994). It is hoped that future work in this area will build upon the methodological guidelines that have been developed and thus minimize the type of problems discussed here.

References

Blackmore, S. 1994. "Probability Misjudgement and Belief in the Paranormal: Is the Theory All Wrong?" In *Proceedings of the 37th Annual Convention of the Parapsychological Association,* edited by D. Bierman, 72–82.

Chapman, J. 2001. "Is There Anybody There? Mediums Perform Well in Scientific Séance Test." *Daily Mail,* March 5.

Hyman, R. 1977. "Cold Reading: How to Convince Strangers that You Know All about Them." *Skeptical Inquirer* 1, no. 2: 18–37.

Matthews, R. 2001. "Spiritualists' Powers Turn Scientists into Believers." *Sunday Telegraph,* March 4.

Milton, J., and R. Wiseman. 1997. *Guidelines for Extrasensory Perception Research.* University of Hertfordshire Press: Hatfield, England.

Rosenthal, R., and D. B. Rubin. 1978. "Interpersonal Expectancy Effects: The First 345 Studies." *Behavioural and Brain Sciences* 3, 377–86.

Rowland, I. 1998. *The Full Facts Book of Cold Reading.* Ian Roland Limited: London, England.

Schouten, S.A. 1994. "An Overview of Quantitatively Evaluated Studies with Mediums and Psychics." *Journal of the American Society for Psychical Research* 88: 221–54.

Schwartz, G. E. R., L. G. S. Russek, L. A. Nelson, and C. Barentsen. 2001. "Accuracy and Replicability of Anomalous After-death Communication across Highly Skilled Mediums." *Journal of the Society for Psychical Research* 65, no. 862: 1–25.

26
NEAR-DEATH EXPERIENCES

ANTONY FLEW

The subject "Near-Death Experiences" is related to the question "Do Souls Exist?" It is related inasmuch as it is thought that the investigation of near-death experiences and what used to be called psychical research might possibly produce evidence of the existence of souls as what philosophers call substances.

Words for substances in this philosophical sense of the word "substance"—unlike, for instance, words for characteristics which have to belong to, to be of something else—are words for something supposed to exist separately and in its own right. This understanding of the meaning of the word "substance" cannot in the present context be more illuminatingly illustrated than by quoting the first definition of the word "death" supplied by Dr. Samuel Johnson in his *Dictionary of the English Language*. It reads: "The extinction of life, the departure of the soul from the body." For Dr. Johnson the soul was just as much a substance, in this philosophical sense, as the body. It was the experiencing, planning, deciding, incorporeal, spiritual substance which had been in charge of the flesh-and-blood person whose body it had been. It is of course

Antony Flew's "Near-Death Experiences" was originally delivered at the conference "Science and Religion: Are They Compatible?" sponsored by the Center for Inquiry, and held in Atlanta, Georgia, in November 2001.

only inasmuch as people are thought of as being essentially their souls, their substantial but incorporeal spirits, that any doctrine of life after the death (of the body) even makes sense.

To this contention, which I have myself been advancing for over fifty years, it has sometimes been objected that the Christian doctrine of a future life involves a resurrection of the flesh. But without an incorporeal but substantial soul to maintain the identity between the since deceased flesh and blood human being and the later resurrected flesh that would be at best a replica and not one and the same human being. If anyone doubts this let them ask themselves whether they would consider that the claims of justice would be satisfied were it possible to construct a perfect, living replica of one of the greatest monsters of the former tormented century—Stalin, Hitler, Mao Tse Tung, and their like—and inflict on them a few multimillennia of confinement in an institution modeled on one of their own political prisons.

The close connection between the questions so far mentioned and the master question of the relation between science and religion can be vividly illustrated by making a literal one or two quotations from an Encyclical Letter from Pope Pius XII, a letter entitled *Humani Generis* (Concerning the Human Genus) and published in 1950.

The first is: "Thus, the teaching of the Church leaves the doctrine of evolution an open question, as long as it confines its speculations to the development, from other living matter already in existence, of the human body." However, "That souls are immediately created by God is a view which the Catholic faith imposes on us." It is, of course, this special creation of every individual soul, introducing a supernatural element into the Nature studied by scientists, which opens up the possibility—in the view of the Church of Rome the certainty—that the essence of each and every human being released from his or her body at death will eventually proceed, freshly equipped with a spiritual body to an eternity of bliss—or, more likely, torture.

My second quotation will surely be more surprising. It reads:

> Christians cannot lend their support to a theory which involves the existence, after Adam's time, of some earthly race of men, truly so called, who were not descended ultimately from him, or else supposes that "Adam" was the name given to some group of our primordial ancestors. It does not appear how such views can be reconciled with the

doctrine of original sin, as this is guaranteed to us by scripture and tradition, and proposed to us by the Church. Original sin is the result of sin committed, in actual historical fact, by an individual man called "Adam," and it is a quality native to all of us, only because it has been handed down by descent from him.

If and when somebody produces a plausible account of how some extremely primitive kind of living and reproducing organism might have evolved from nonliving matter, then the embargo on such evolutionary speculation could surely be lifted without an intolerable degree of embarrassment for the Vatican. But this insistence upon the catastrophic consequences "of a sin, committed, in actual historical fact, by an individual man named Adam" is explicitly justified by an appeal to the Decree concerning Justification issued by the Fifth Session of the Counter-Reformation Council of Trent. The repudiation of such a decree from that of all Councils would not but be seen as the affective abandonment of the claim of the Church of Rome to supreme authority in matters of religion and morals.

It is sometimes suggested that Out-of-the-Body experiences (OBEs) provide evidence for the existence of a substantial (in that philosophers' sense) but of course immaterial and hence spiritual soul. I myself know of no one who denies that such experiences are commonly had, often by patients recovering from periods of apparently deep unconsciousness. Those, however, who challenge their significance as phenomena supposedly tending to confirm some survival hypothesis are often accused of such blinkered denials. But the truth is that the only OBEs of the kind which would have any claim to possess a cognitive status higher than that of any other dream would be those which apparently contained information not normally available to the patient in question. Such patients were typically, at the times when they were enjoying their OBEs, lying apparently unconscious on their hospital beds. So if OBEs of this apparently cognitive kind do actually occur, then this ought to be taken as evidence for the reality of Extrasensory Perception (ESP) rather than as evidence for the making of explorations by some temporarily disembodied soul. For, even if sense can be given to the idea of a disembodied soul, the postulation of the involvement of such an incorporeal explorer would be grossly uneconomical. For having no body, and consequently no sense organs, any information brought back from its alleged travels would have to have been acquired by ESP.

So what about Near-Death Experiences (NDEs)? The classic objection to offering NDEs as evidence for the existence of a substantial soul separable from the patient's body is that such experiences are all experiences reported by—surprise! surprise!—still living persons. This objection to such evidence can be overcome only if the experiences reported by the of course still living patients include information which could only have been acquired, presumably by ESP, at times when there were no goings-on in the brains of those patients either to be or to be the cause of their consciousness.

The only case I know which even appears to satisfy this requirement is one to which I drew attention in a recent issue of the *Skeptical Inquirer*.[1] It was one recorded by Dr. Michael Sabom in his *Light and Death: One Doctor's Fascinating Account of Near-Death Experiences* (Grand Rapids, Michigan; Zondervan, 1998). A patient, Pam, is there alleged to have provided, after her recovery, a blow by blow account of the activities of the surgical team at times when she was reputedly brain-dead. But this case of an alleged NDE apparently depends upon the unsupported testimony of the author which, in a matter of such eschatological importance, is quite scandalously inadequate.

In a letter in a later issue of the same journal (November/December 2001) Dr. Michael Sabom responded to this objection by insisting that his account of the surgical procedures was supported by the "world-renowned surgeon" who conducted the entire operation. But, significantly, he refrained from claiming that that surgeon was himself involved in the "structured interview protocol" by which "Details of Pam's NDE were obtained" and is himself prepared to assure us of the reliability of Dr. Sabom's account.

In October 2000 there were stories in the British press of an investigation in the Southampton General Hospital by Doctors Sam Parnia and Peter Fenwick into the supposed recollections of survivors of cardiac arrest. These recollections were welcomed by some bishops as evidence of death survival. But after the report of "A qualitative and quantitative study of the incidence, features, and aetiology of near-death experiences in cardiac-arrest survivors" was actually published—in the January 2001 issue of *Resuscitation*[2]—it became obvious that they had no such significance.

For in the British journal the *Skeptic*[3] the associate editor, who had been able not only to read the report but also to discuss it with Dr. Peter Fenwick, one of its authors, provided a devastating summary of its deficiencies for our present purposes. In the first place there had been no direct measurement of brain activity or the absence of it during the periods of cardiac arrest. In the second place, since a cardiac patient loses consciousness fairly rapidly and (if lucky!) recovers it somewhat more slowly it would clearly be impossible to attribute any experiences recorded up to a week later to the specific period when there was no brain activity. In the third place there is the strong possibility that a brain which has had a spell of shutdown (brain death) might generate some false memories to fill the gap. Only if the putative memories of the patients embraced information about the circumstances surrounding them during the periods of their brain deaths, and only if they had no normal means of acquiring this information before they reported these putative memories, would we be warranted in describing these cases as either cases of retrospective ESP and/or cases of experiences had by the substantial souls of the patients.

The editor of the *Skeptic* concluded his review of this case tersely: "In summary . . . the Southampton research showed that out of 63 cardiac arrest survivors, four exhibited some of the (subjective) attributes of NDEs. Not much to write home to mother about, really!" A much more general conclusion was drawn by Susan Blackmore in her recent article on "What Can the Paranormal Teach Us about Consciousness" (*Skeptical Inquirer*, March/April 2001): "The more we look into the workings of the brain the less it looks like a machine run by a conscious self and the more it seems capable of getting on without one."

Notes

1. Antony Flew, "Letters to the Editor," *Skeptical Inquirer* 25, no. 4 (July/August 2001).

2. Peter Fenwick and Sam Parma, "A Qualitative and Quantitative Study of the Incidence, Features and Aetiology of Near Death Experiences in Cardiac Arrest Survivors," *Resuscitation* 48, no. 2 (2001).

3. Steve Donnelly, "Rhyme and Reason," *Skeptic* (UK) 14, no. 2 (2001).

27
DOES THE SOUL EXIST?
From the Mythological Soul to the Conscious Brain

JEROME W. ELBERT

I'll start by giving a "generic" definition of the soul that includes some soul beliefs from other cultures. I define a soul as, "A very special part of a human being, in addition to the body, that gives a person at least one of the following: life, a personality; or, the ability to move oneself, to think and feel, to leave the body, to know right from wrong, to survive death and perhaps be reincarnated, to exercise free will, or to have a spiritual relationship with God."

I have only defined souls for people, but many cultures hold that animals also have souls. In some cultures, the different powers of the soul are contained within various souls, so someone can have a number of souls that provide the person with such different things as life, a personality, and immortality.

The Belief in Souls Is Ancient and Widespread

Throughout history, most people believed in souls. In certain rural folktales told from Ireland to Japan, a person falls asleep, and a small animal or insect

Jerome W. Elbert's "Does the Soul Exist? From the Mythological Soul to the Conscious Brain" was originally delivered at the conference "Science and Religion: Are They Compatible?" sponsored by the Center for Inquiry, and held in Atlanta, Georgia, in November 2001.

leaves the person's mouth and roams about. Later the small organism returns, crawls into the mouth of the sleeping person, and the person awakens. Sometimes the stories imply that the person will die if the small creature is prevented from returning into the person. Sometimes the adventures of the small animal outside the person's body correspond to an unusual dream experienced by the sleeping person (Bremmer 1983, 64–66, 132–35). The common features of these tales suggest that prehistoric soul beliefs may have diffused throughout Europe, Asia, and presumably Africa.

There is also evidence for prehistoric soul beliefs in Australia. (For example, see Swain 1993, 100, 101). The native Australians, or Aborigines, arrived in Australia at least 40,000 years ago. After arriving, they were largely isolated from outside influences. They did not adopt such "modern" inventions as bows and arrows or agriculture. But they believed in souls. They believed that, after death, their souls went away to a spirit-place. Some aborigines believed that these souls could be reused by new members of their own clan. These beliefs, present when Europeans arrived in Australia, suggest that the Aborigines already believed in souls, or had a tendency to believe in souls, when they *came* to Australia 40,000 or more years ago. If this conjecture is correct, the idea of souls is tens of thousands of years old. A similar argument, over a time interval of about 12,000 years, can be made for native Americans.

Soul beliefs varied a great deal in early recorded history. The Hebrews, Persians, and Greeks contributed to our modern ideas about the soul. Before the 6th century B.C.E., the Hebrews, along with the Phoenicians, Babylonians, Greeks, and Romans, believed that the soul barely existed after death. The soul might survive as a shadowy entity, or "shade" in the underworld, but it was supposed to have only a dull existence without clear consciousness. Concerning the early Greek ideas about souls, Bremmer said that, "On the whole they are witless shades who lack precisely those qualities that make up an individual" (Bremmer 1983, 124).

Perhaps Zoroaster was the first person who had ideas about the soul that were close to modern ideas. He was a Persian who may have lived about 600 B.C.E., or earlier. In his religion, he taught that humans have a body and a soul, and that *the soul gives us the faculties of reason, consciousness, conscience, and free will.*

According to Zoroastrianism, free will enables people to choose good or evil actions and makes them morally responsible for their actions. Since people make their own choices, God is justified in rewarding or punishing them by sending them to heaven or hell. About 1,000 years later, the extremely influential Christian theologian Augustine made similar arguments. Augustine may have been indirectly inspired by Zoroaster, since, for nine years before he became a Christian, he was a Manichean. That sect had been influenced by Zoroastrian and other beliefs.

Although the early Hebrews and Greeks had not emphasized life after death, many changed their minds later, at least partly under the influence of the Persians. Some Jews, during difficult times of political oppression in which justice on earth appeared to be absent, began to believe in life after death. Following the Babylonian Exile, the Jews had very close contact with the Persians. At that time, and later, many of the Jews adopted beliefs similar to those of the Zoroastrians, such as the resurrection of the dead, a final judgment, and reward and punishment by God after the judgment. Previously, these had not been typical Jewish beliefs. In the adopted picture, eternal life was a reward of the just, and the Hebrew underworld became a place of punishment for sinners (Elbert 2000, 64–67).

For some time, there were disagreements among the Jews, between the *Sadducees,* who held the older beliefs, and the *Pharisees,* who accepted the Persian beliefs. By the start of the Christian era, however, the Persian beliefs were widely accepted by Jews. At that time, Jewish beliefs about an afterlife emphasized a resurrected body united with a soul, with a fully human existence.

Early Greek ideas about the soul can be inferred from the ancient writings attributed to Homer. *Psyche* was the most important Greek soul-word. In Homer, the psyche leaves a person who faints or dies. Other soul-words were used to describe the sources of a person's psychological traits and most mental abilities. At death, only one's psyche survived (Elbert 2000, 39, 40). It had no emotions, drive, or ability to think. So, a dead person became the kind of "witless shade" I mentioned earlier.

Later, when science and philosophy started to grow in Greece, many different ideas about souls developed. Thales, the founder of Greek philosophy, believed that any object that moved itself under its own power showed evidence that it had a soul. Since a magnet can move itself toward a piece of iron, Thales thought magnets have souls!

Many Greek philosophers before Socrates tried to understand everything in terms of matter and its interactions. Democritus, who introduced the idea of atoms, proposed that the soul is made of very mobile spherical atoms.

Others thought that the soul was a gas or liquid. There was little evidence to support such explanations of the soul. The weakness of these early attempts to explain the soul in natural terms allowed occult or spiritual explanations of the soul to be proposed by others.

The early Greek cult of Orphism taught that a person is a combination of a soul of divine origin and a body of a much lower nature. After death, a person's soul was judged and sent to *Elysium* (similar to heaven) or to hell. After a period of reward or punishment, the soul returned to earth in another body. The cycle was repeated a number of times, but the soul eventually returned to its divine source. Pythagoras, a mathematician and cult founder, had similar beliefs.

The great philosophers Socrates, Plato, and Aristotle influenced later Christian soul beliefs. During the fifth and fourth century B.C.E., Socrates greatly increased the importance of the soul by treating it, not the body, as the real person. This was a radical idea for most Greeks of that time.

Plato, perhaps influenced by Orphism and Pythagoras, maintained that the soul is spiritual, immortal, and has no parts. He believed the soul is the source of one's mental activities and that it causes the body's movements. He regarded the soul as superior to the body and the most important part of a person. His ideas about the soul had a great influence on early Christian theologians.

Unlike Plato, Aristotle tended to describe humans in natural rather than supernatural terms. For Aristotle, body and soul are more of a unity, just as "the wax and the shape given to it by the stamp are one."

Now let's shift to Christian beliefs about the soul. Presumably, very early Christian ideas about the soul were strongly influenced by some Jewish beliefs that were current at that time. One indication of very early Christian beliefs is a statement by Jesus in the Gospel of Matthew, chapter 10, verse 28. He says, "And do not be afraid of those who kill the body but cannot kill the soul; but rather be afraid of him who is able to destroy both the body and soul in hell." This suggests that the soul survives death and that the body will eventually be reunited with the soul.

Most likely, there were great variations in what early Christians believed about the soul. Certainly, early theologians disagreed on soul beliefs. Tertullian, who lived from about the year 160 to about 220 C.E., held that the soul is *physical* and develops with the body. Origen, who lived a little later, had studied Plato's ideas. Origen held that the soul has a spiritual nature and that it is immortal, and that it exists *before* the body. Jerome, who lived from about 345 to 420, argued that the soul is created at the time of conception. This claim still influences some modern debates on the issue of abortion.

Augustine, one of the most influential theologians in Christian history, lived from 354 to 430. His views dominated Christian soul beliefs for about eight hundred years, from his lifetime to well into the Middle Ages. His thinking was strongly affected by Plato's ideas. According to Augustine, a human being is an immortal, spiritual soul *using* a mortal body. To Augustine, the soul was far superior to the body. Augustine's teachings form a link between the beliefs of Plato and those of some present-day Christians.

Much later, around the thirteenth century, Thomas Aquinas picked up Aristotle's ideas and introduced a more unified picture of our basic nature into western Europe. That is, he taught that a person is more like a single entity than the easily separable body and soul advocated by Augustine and Plato. After about a generation of debate, the Catholic Church largely accepted Aquinas's teachings about the soul. For everyday soul beliefs, however, Aquinas did not have the last word.

In the seventeenth century René Descartes, a French mathematician and philosopher, shifted people's thinking back toward Plato's views about a dual human nature. That is, Descartes persuasively argued that we have two parts: an immaterial mind, or soul, and a physical body. *Although his arguments have lost their appeal to most modern philosophers and scientists, his ideas survive in everyday speech and thought.* Descartes's dualistic picture of our nature, consisting of a body and a soul, was reasonably satisfactory to the Church and, at that time, to the emerging sciences

Descartes changed the way we think about ourselves by highlighting *consciousness* as the most important aspect of mind. His writings linked the "soul" and the "mind," since both are supposed to support consciousness. *Since Descartes, the mind has taken over many of the soul's mysterious characteristics.* This tends to inject an element of mystery into our feelings about consciousness.

Soul beliefs are not just of historical interest. A few years ago, *Time* magazine reported on a poll of American adults which asked whether souls exist. An overwhelming 92 percent reported that they believe in the soul. Many present-day Hindus, Moslems, Christians, and Jews believe in some sort of a supernatural soul.

The Soul and Two Conflicting Worldviews

There are opposing views on the existence of souls. I will describe this disagreement as a conflict between two worldviews: the traditional and the scientific.

In the traditional worldview, matter has very little potential, like a stone that keeps its shape and stays in one location. By itself, matter, sometimes described as "mere matter," is not supposed to be capable of living or thinking. For matter to exhibit life and mental abilities, something must be added from outside the material world. A soul is supposed to fill this need, making it possible for the human body to be alive and to think. Just how the soul does this is held to be a mystery.

As a bonus for soul believers, the idea of a soul suggests that immortality is possible. The soul is not made of visible matter, so no one has ever seen a soul decay. It is claimed that the soul has no parts, so how could it fall apart? So, it is conceivable that the soul remains intact after one's death. Since consciousness is a property of the soul in this worldview, it is suggested that humans can go on thinking and feeling after death because the soul survives.

The Original Sin of the Mind

The popularity of the idea of the immortal soul is understandable. It is not surprising that people enjoy thinking about schemes that promise life and happiness after death. By a process which I will call *the original sin of the mind*, people tend to accept such attractive ideas, even though evidence for them is lacking. Such ideas have controlled nations and empires, and perhaps the majority of the world's people hold the traditional ideas about souls.

The Traditional Worldview Is Deeply Embedded

The traditional worldview is so deeply embedded in our culture that it often seems intuitively correct. For example, it is ironic that we will admit that every mental ability can be destroyed by damage to the brain, but we tend to doubt that all of our mental abilities arise from the ordinary matter that makes up our brains. We may admit that our feelings can be altered in many ways by drugs, but we are hesitant to believe that naturally occurring brain chemicals can explain why we feel the way we do. We often feel that something else is needed. As another example, even a person without religious faith may believe that free will allows people to perform minor miracles in making personal decisions, or that "mere matter" cannot account for our beliefs, values, or feelings.

The Scientific Worldview

Now let's discuss souls from the perspective of the scientific worldview. It emphasizes experimental and observational evidence rather than tradition and intuition. It avoids making claims that are not supported by strong empirical or observational evidence. It is open to all sorts of hypotheses, but does not accept these hypotheses without proof.

In recent centuries, the scientific worldview has made great progress in providing us with a new understanding of the physical and biological world. Because this scientific knowledge rests on a solid foundation, the same body of knowledge is accepted by scientists in highly diverse cultures in all parts of the world.

Scientific research has shown that matter is surprisingly capable of displaying wonderful properties. A small number of types of fundamental particles, with a few kinds of basic interactions between them, seem to account for all physical and biological phenomena. Science has found no evidence of supernatural entities or occurrences and, of course, it does not claim that any exist.

Science's case against the soul is stronger than just a lack of convincing evidence favoring the idea of the soul. The scientific perspective *removes the underlying motivations* for using the soul to explain life, our mental properties, immortality, and free will. I will briefly summarize this situation.

Scientific progress in understanding nature has led to many cases where natural explanations replaced earlier traditional beliefs that depended on supernatural processes. Prominent examples include the origin of the solar system, the development of animal structures and functions by biological evolution, and the inheritance of these properties through DNA. Because of the great progress made in biology, *life* no longer seems to require a supernatural explanation, such as the presence of a soul.

Brain studies have reached the point where most neuroscientists believe that all mental processes will be understood in terms of the functioning of the brain. Thus, according to most experts, our mental abilities do not require the existence of a supernatural soul.

The scientific understanding of life and human consciousness implies that death is the end of conscious life, since it involves the end of activity in the brain. From this perspective, the idea of immortality is mistaken.

Consequently, if the idea of immortality is wrong, there is no need for a soul to make us immortal.

How about the soul as the source of free will? Human decision making is now considered to be a physical process in the brain. As such, it should not involve the sort of "minor miracles," or "microscopic spoon-bending" implied by some traditional beliefs about free will. If our decisions are entirely natural results of brain processes, our freedom of choice does not require a supernatural soul (Elbert 2000, 297–330). In this approach, quantum processes are not crucial to the free-will debate, since there is no reason to believe that these *random* effects are controlled by brain processes.

Near-death experiences are a modern reason why some people believe in the soul. Perhaps near-death experiences even influenced ancient ideas about the existence of souls. I believe that these phenomena are understandable as exceptional experiences produced by brains in very stressful circumstances. This position, supported, for example, by the book *Dying to Live: Near-Death Experiences* (Blackmore 1993), is that the experiences arise within a person's brain as a result of such physiological processes as oxygen deprivation and the release of natural opiates. From this perspective, the experiences are entirely dreamlike or hallucinatory, and they are not the result of actual interactions with departed relatives or well-known religious figures. They do not prove the existence of supernatural souls.

The idea that a soul allows one to have a spiritual relationship with God implies that an imagined relationship with God is unique, and somehow different from the range of relationships with other imaginary entities. As far as science is concerned, there is no reason to believe this is true. There is, of course, no reason to believe a soul is needed to carry out this dubious role.

The straightforward conclusion of the scientific worldview is that there is no convincing evidence or reason to believe in souls. Without strong evidence or good reasons to believe in them, Occam's razor dictates that we reject the idea of supernatural souls.

Two extremely well-known scientists have spoken out against the idea of the soul. Francis Crick, renowned for his early work on the structure of DNA, has since worked on consciousness and the brain. He says, "A modern neurobiologist sees no need for the religious concept of a soul to explain the behavior of humans and other animals."

In 1921 Albert Einstein wrote that "the concept of a soul without a body seems to me empty and devoid of meaning."

Science-Based Objections to the Traditional Soul

There are other reasons why the idea of a supernatural soul is contrary to the scientific perspective. If souls exist and are essential for thinking and decision making, our mental processes involve frequent communications from the brain to the soul and from the soul to the brain.

As a scientist, I find the idea of such interactions very disturbing. If such interactions exist, the human brain is an interface to another, nonphysical world. Such interactions suggest that the rules of science apply to all the universe except for human beings. We, however, supposedly interact continually with this other world, and this other world continually affects our behavior. This picture gives humans a unique position in the universe. This anthropocentric picture seems very unacceptable to the scientific worldview.

The idea that human bodies constantly interact with minds that are not part of the physical world is also contrary to the fundamental belief held by physicists that all physical processes are based upon fundamental interactions between subatomic particles. Suppose we grant that any description of a human being must make sense when viewed on the microscopic scale of the interactions between the elementary particles that make up the human being. We immediately run into a problem, since physics knows of no interactions between the physical world and some separate spiritual world. Clearly, if elementary particles are affected by interactions with a spiritual world, a major revolution is coming in physics. Based on the lack of convincing evidence for the soul, however, there is little reason to expect it.

I think it is amusing to think of how a supernatural soul might affect human evolution. If our thinking can be done outside the body, we might expect evolution to produce humans with brains that, over thousands of years, would become smaller and smaller. That is, if the really difficult thought processes can be handled outside the physical world, then human brains may eventually be reduced to mere neural gateways, or interfaces, to a spirit world. As far as I know, however, no supporter of the soul has proposed such a bizarre outcome.

Complexity and the Soul

Modern technical knowledge implies that anything that supports our mental abilities must be a very complicated thing. We have memories that store thousands of gigabytes of information. It is reasonable to think that the brain, with perhaps a million billion (10^{15}) neural connections, can contain this information. It is not reasonable to think that a simple entity, such as a soul is often supposed to be, could carry on normal mental abilities, such as consciousness, after death.

Immortality without the Soul

Some Christians do not believe in an immortal soul, but hold that we are immortal because of the biblical promise of the resurrection of the dead, a final judgment, and reward or punishment in heaven or hell. These people tend to believe this was revealed by God to the Bible's writers. There are many reasons not to accept the Bible's word on this. In particular, there is a good reason to be skeptical, since all of these ideas were held by the Zoroastrians of Persia before they were widely accepted by Jews. Christianity may have received these beliefs thirdhand. A skeptic is justified in believing that this kind of immortality belief is ultimately based on pagan myths, wishful thinking, and various accidents of history, not divine revelation.

Topics Where Soul Views Make a Difference

The idea of the soul can have effects in the real world, even if supernatural souls don't exist. The Roman Catholic Church teaches that human rights start abruptly at conception, since an individual's soul is supposed to be created at that time. Members of other religions and agnostics, on the other hand, may believe that it is ethical to destroy an embryo or fetus before neural activity begins in the brain. At this early stage, there has been no mental activity, and no person has developed.

Sometimes, Catholics and religious conservatives support political positions on early abortions and on stem cell research that are motivated primarily by theology. Although the United States proclaims the separation of church and state, our government has sometimes adopted laws and policies on these topics that are motivated primarily by the supernatural beliefs of certain denominations.

Consider cloning. I concede that, at this time, it may not be safe to use cloning to *reproduce* humans. A *different proposal*, to use cloning to produce clusters of about a hundred cells for medical research, has been opposed by some religious groups. The opposition to this kind of cloning is heavily based on religious beliefs. Blocking this medical research comes close to elevating certain supernatural beliefs to the level of national doctrine. Blocking this research is against the public interest, since people might benefit from medical advances based on the proposed work.

Separate Domains and the Soul

Some people maintain that science and religion are compatible because they involve entirely separate domains. Soul beliefs seem to be a counterexample to that position. Long ago, when religions adopted soul beliefs, they unwittingly trespassed into a domain that has implications for science. At present, science does not support, and tends to contradict, the traditional religious positions regarding the soul.

Souls, Persons, and a Personal God

If one considers the possibility of a supreme being similar to the God of Judaism, Christianity, and Islam, one may ask why this being would be a *person* and why this being would be a *spirit*. Although I may seem to be straying from the main topic, I will show the connection between this issue and the previous discussion of souls.

First let's consider why God might be a spirit. Perhaps this is because God is traditionally regarded as the creator of all material things. This suggests that God is "above" and "outside" the material world. Also, since God is not supposed to be mortal, while at least the bodies of humans are mortal, it seems reasonable to believe that such a God would be outside the material world.

From the scientific perspective, our properties and abilities as *persons* are entirely natural in origin. We possess these properties and abilities because of our evolutionary background. Humans tend to have these characteristics because they have survival value. By accident and natural selection, our species acquired the characteristics that we bundle into the idea of a "person."

Now let's consider why God would be a person. If one supposes that a supreme being is the creator of all things, this being would not be a person for the same reason that we are persons. We are persons because of evolutionary processes, but natural selection only works on *reproducing, mortal* organisms. The traditional God of the Western religions is not in this category.

We must look elsewhere to understand why God would be a person. It helps if we view the idea of God skeptically, as a product of the traditional worldview, not the scientific worldview. In that worldview, people are not persons because of natural processes. They are persons because they are made in the image of God. Of course, that still does not explain why God is a person. From a skeptical perspective, God is a person because some ancient thinkers created God in their own image. This was the easiest way to imagine an immortal supreme being: as a glorified human, but without the mortal body. *The idea that God is a spirit was supported by the mistaken idea that the highest part of human nature is spiritual.* This was nicely consistent with the reasons given above, in the second paragraph of this section, for why the traditional God would be a spirit. As I have argued on previous pages, however, the scientific perspective does not support the idea that any part of human nature is spiritual. As argued here, the scientific perspective also does not support the idea that God is a person (*also see* Elbert 2000, 345–48). Without a nervous system, or physical equivalent, there is no reason to believe that God would be conscious. Stated simply, the scientific perspective does not support the traditional God.

Conclusions and Final Remarks

The body of knowledge assembled by science not only does not support the idea that human beings have supernatural souls, but it removes the basic motivations for believing in souls. I have also argued that current physical theories are not consistent with the idea of supernatural souls, since the theories do not include any interactions between the material world and any other world. Such interactions would be needed to account for the human mind, conscience, or free will if they are powers of an immaterial soul. Also, contrary to some soul beliefs, science does not, of course, suggest that humans are immortal.

I have argued that science and the idea of a *supernatural soul* are quite incompatible. There is no inconsistency with science, however, in defining an entirely *natural soul* that consists of a bundle of ordinary human abilities such as one's consciousness, will, conscience, feelings, and emotions. I discussed this in the section "Introducing the Natural Soul" in *Are Souls Real?* (Elbert 2000, 268–69).

In conclusion, the scientific perspective does not support the view that humans have a partly immaterial or spiritual nature. In fact, I have argued that the scientific perspective does not support the idea that there are personal, spiritual beings: human or divine. Although some people succeed in holding both scientific and religious views, it is difficult to reconcile the two. Perhaps as a consequence of this, the percentage of American scientists who are religious is much smaller than the corresponding fraction of all Americans.

References

Blackmore, Susan. 1993. *Dying to Live: Near-Death Experiences.* Amherst, N.Y.: Prometheus Books.

Bremmer, Jan N. 1983. *The Early Greek Concept of the Soul.* Princeton, N.J.: Princeton University Press.

Elbert, Jerome W. 2000. *Are Souls Real?* Amherst, N.Y.: Prometheus Books.

Swain, Tony. 1993. *A Place for Strangers: Towards a History of Australian Aboriginal Being.* Cambridge, UK: Cambridge University Press.

28
EFFICACY OF PRAYER

IRWIN TESSMAN AND JACK TESSMAN

The therapeutic power of prayer is a recurring theme among many proponents of alternative medicine. One can imagine a natural explanation for the alleged benefits: a psychological boost from the belief that a supernatural power is on your side. But what if you are unaware that people are praying for you? Such intercessory prayers could only work through a supernatural agency.

Investigating the efficacy of intercessory prayer was given scientific legitimacy by Francis Galton, the father of biometry and a central figure in the founding of modern statistical analysis. In classic memoirs, Galton (1872, 1883) argued that regardless of how the prayers "may be supposed to operate," the "efficacy of prayer . . . is a perfectly appropriate and legitimate subject of scientific inquiry" because it can be tested statistically, as he then demonstrated.[1]

A Landmark Study

A celebrated study performed at San Francisco General Hospital by Randolph C. Byrd reported that patients in a cardiac care unit received statistically significant benefits from intercessory prayers (Byrd 1988). That study has

Irwin and Jack Tessman's "Efficacy of Prayer" originally appeared in the *Skeptical Inquirer* 24, no. 2 (March/April 2000).

attained special status within the alternative medicine community and has been reprinted as a "landmark study" (Byrd 1997).

In the same skeptical spirit that motivates one to seek the flaw in the design of a perpetual motion machine, we have examined Byrd's study, as others have done (Posner 1990; Sloan et al. 1999; Witmer and Zimmerman 1991), to seek a natural explanation to rival the supernatural one.

We believe a serious flaw exists in his critical Table 3, a flaw that raises doubts about the table's validity. That table reports the overall outcome for patients admitted to the cardiac care unit. Upon admission, they were entered randomly into one of two groups: an intercessory prayer group or a control group (192 and 201 patients). The outcome was recorded as good, intermediate, or bad.[2] Byrd found that compared to the control group, the prayer group had an excess of good outcomes and a deficit of bad outcomes, a significant difference in favor of the prayer group with $P < 0.01$ (Byrd's Table 3). The study was necessarily intended to be double blind. Byrd writes: "The patients, the staff, and doctors in the unit, and I remained 'blinded' throughout the study."

Unfortunately, that was not the case at a critical point. Byrd's Table 3, which might best have been constructed by a panel of "blinded" doctors, was constructed by Byrd alone. But it was done in response to criticism of an earlier version of his manuscript, the writing of which had already required that the code be broken. Thus Byrd was no longer blinded when he determined the answer to the key question of which did better, the intercessory prayer group or the control group (Byrd, personal communication).

Because the table was apparently constructed from computer-stored data using objectively stated criteria that did not involve Byrd in any personal evaluation of individual cases, the lack of blinding might have had no effect. Although blind evaluation is clearly preferable, the use of an unblinded analysis could be defended were it completely computer generated. However, the criteria he chose for evaluating the patients' outcomes were formulated after the data were collected and when Byrd was unblinded. That is an unreliable approach. The criteria should have been selected before the start of the study.

The claim of blindedness is erroneous in yet another respect (one aspect of which has already been mentioned [Witmer and Zimmerman 1991]). In his acknowledgments, Byrd thanks "Mrs. Janet Greene for her dedication to this study," but without any elaboration of her role. In a later publication (Byrd with Sherrill 1995) we learn that Janet Greene was hired ". . . to be our coordinator.

. . . Janet entered names of all the volunteer patients into a computer that randomly divided them into two groups. . . . half of the patients—only Janet knew who they were—were prayed for daily by our intercessors. . . . She kept detailed records of all patients in both groups." Thus the very coordinator of the study was completely unblinded. Once patients were assigned to one of the two groups, Greene should have had no further contact with the hospital.[3]

Byrd's evidence for supernatural intervention, if true, would arguably be one of the most remarkable scientific demonstrations of the last millennium. To be credible, however, it requires, among other things (Posner 1990; Sloan et al. 1999; Witmer and Zimmerman 1991), considerably more attention to strict blindedness. In the absence of that credibility, its status, not to mention the "landmark" label, is highly dubious.

A Confirmation Attempt

Recently, another prayer study, broadly based on Byrd's (and the subject of numerous news reports in October and November 1999), examined 990 patients admitted to a coronary care unit (Harris et al. 1999).[4] The authors scored the effects of intercessory prayer on the occurrence of thirty-four adverse conditions (Harris's Table 3).[5] These are similar to the twenty-six conditions scored by Byrd (his Table 2).

Their general approach to scoring the efficacy of intercessory prayer is summarized as follows: "Since prayer was offered for a speedy recovery with no complications, it was anticipated that the effect of prayer was unlikely to be evident in any specific clinical outcome category (e.g., the need for antibiotics, the development of pneumonia, or the extension of infarction), but would only be seen in some type of global score."

Let us therefore look first at the speed of recovery. The length of stay in the coronary care unit decreased 9 percent in the prayer group, but with P = 0.28;[6] the length of hospital stay increased by 9 percent in the prayer group, but with P = 0.41 (their Table 4). Thus, by either measure the large P values indicate that the results are quite consistent with a null effect; thus there is no evidence that intercessory prayer confers any benefit (or harm) in speed of recovery.

Next we examine the results for two types of global scores. One is the Mid America Heart Institute–Cardiac Care Unit (MAHI-CCU) weighted score (their Table 4)[7] for the thirty-four adverse conditions. They call this score the "primary predefined end point" of their study. It shows an 11 percent advantage to the intercessory prayer group with P = 0.04.

Another type of global score arises from an evaluation of overall outcomes judged by a blinded panel to be either good, intermediate, or bad, each based on Byrd's criteria. Whereas Byrd found a significant difference (P < 0.01) in good and bad outcomes in favor of the prayer group, Harris et al., using the same criteria, find no significant difference (P = 0.29, Harris's Table 5). Thus, not only do these results of Harris et al. fail to confirm the significant differences found by Byrd, they constitute a second set of results (the first being on speed of recovery) that shows no significant effects of intercessory prayer.

Thus Harris et al. make three major tests of the efficacy of intercessory prayer: speed of recovery scores (Table 4), MAHI-CCU global scores (Table 4), and outcome scores (Table 5). On the basis of just the MAHI-CCU scores taken alone with its barely significant P = 0.04 value, Harris et al. conclude there is a beneficial effect of intercessory prayer.

This argument is simply fallacious: where there are multiple tests it is incorrect to single out just one, ignoring others with large P values that indicate no significant differences between the groups tested. For example, if the three tests were completely independent, the probability that at least one of the three would show P = 0.04 purely by chance would be $1 - 0.96^3 \cong 0.12$, which is well above the conventional maximum value of 0.05 for significance. Though the tests are not independent, it is clear that the overall probability of observing that just one of these three tests favors intercessory prayer with P as low as 0.04 is well explained by pure chance.[8]

Conclusions

The tests of Harris et al., taken in their entirety, fail to show any significant benefit of intercessory prayer, and one of the tests directly contradicts Byrd's primary evidence for efficacy (his Table 3) that is the cornerstone of his "landmark study."

Acknowledgment

We thank Louis J. Cote (Purdue University) for extensive discussion and criticism.

Notes

1. Galton's retrospective analysis revealed no beneficial effect.

2. Here is how good and bad scores were achieved. The outcome was scored as good if only one of the following occurred: "left heart catheterization; mild unstable angina pectoris of less than six hours' duration; self-limiting ventricular tachycardia within the first seventy-two hours of myocardial infarction; supraventricular tachyarrhythmia; uncomplicated third-degree heart block requiring temporary pacemaker; mild congestive heart failure without pulmonary edema; no complications at all." The outcome was scored as bad if there occurred "nonelective cardiac surgery, readmission to the coronary care unit after a myocardial infarction with unstable angina, extension of initial infarction, cerebrovascular accident, cardiopulmonary arrest, need for artificial ventilator, severe congestive heart failure with pulmonary edema and pneumonia, hemodynamic shock due to sepsis or left ventricular failure, death."

3. Byrd might have gone further and designed his study so that no human would know, until the appointed time for breaking the code, which patients were in the test group and which in the control group. If the intercessors needed names for their assigned patients, pseudonyms could have been used without any human knowing to whom the pseudonyms referred. This should present no difficulty to the Judeo-Christian God to whom the intercessors were praying.

4. To help assure blindedness, not even the patients knew they were being studied. The requirement of informed consent was waived, in part, because it was felt that the study posed no known risk to either patient group.

5. The conditions include, for example, the need for antianginal agents, antibiotics, arterial monitor, vasodilation, antiarrhythmics, catheterization, diuretics, a permanent pacemaker, an interventional coronary procedure, intubation/ventilation, major surgery, and twenty-two others.

6. We calculate $P = 0.36$. Reminder: P is the probability of this result occurring purely by chance. Conventionally, a value of P greater than 0.05 attributes no statistical significance to the result.

7. An example of their scoring system: if "a patient developed unstable angina (1 point), was treated with antianginal agents (1 point), was sent for heart catheterization (1 point), underwent unsuccessful revascularization by percutaneous transluminal coronary angioplasty (3 points), and went on to coronary artery bypass graft surgery (4 points), his weighted MAHI-CCU score would be 10."

8. It is the responsibility of Harris et al. to calculate the overall P value.

References

Byrd, R. C. 1988. "Positive Therapeutic Effects of Intercessory Prayer in a Coronary Care Unit Population." *Southern Medical Journal* 81: 826–29.

———. 1997. "Positive Therapeutic Effects of Intercessory Prayer in a Coronary Care Unit Population." *Alternative Therapies in Health and Medicine* 3, no. 6 (November): 87–91.

Byrd, R. C., with J. Sherrill. 1995. "The Therapeutic Effects of Intercessory Prayer." *Journal of Christian Nursing* 12, no. 1: 21–23.

Galton, F. 1872. "Statistical Inquiries into the Efficacy of Prayer." *Fortnightly Review* n.s., 12, 125–35.

———. 1883. *Inquiries into Human Faculty and Its Development.* New York: Macmillan and Co., 277–94.

Harris, W. S., M. Gowda, J. W. Kolb, C. P. Strychacz, J. L. Vacek, P. G. Jones, A. Forker, J. H. O'Keefe, and B. D. McCallister. 1999. "A Randomized, Controlled Trial of the Effects of Remote, Intercessory Prayer on Outcomes in Patients Admitted to the Coronary Care Unit." *Archives of Internal Medicine* 159: 2273–78.

Posner, G. P. 1990. "God in the CCU?" *Free Inquiry* 10, no. 2 (spring): 44–45.

Sloan, R. P., E. Bagiella, and T. Powell. 1999. "Religion, Spirituality, and Medicine." *Lancet* 353: 664–67.

Witmer, J., and M. Zimmerman. 1991. "Intercessory Prayer as Medical Treatment? An Inquiry." *Skeptical Inquirer* 15, no. 2: 177–80.

29
SCIENCE VERSUS
SHROUD SCIENCE

JOE NICKELL

The Shroud of Turin continues to be the subject of media presentations treating it as so mysterious as to imply a supernatural origin. One recent study (Binga 2001) found only ten scientifically credible skeptical books on the topic versus over 400 promoting the cloth as the authentic, or potentially authentic, winding sheet of Jesus—including most recently a revisionist tome, *The Resurrection of the Shroud* (Antonacci 2000). Yet since the cloth appeared in the middle of the fourteenth century it has been at the center of scandal, exposés, and controversy—a dubious legacy for what is purported to be the most holy relic in Christendom.

Faked Shrouds

There have been numerous "true" shrouds of Jesus—along with vials of his mother's breast milk, hay from the manger in which he was born, and countless relics of his crucifixion—but the Turin cloth uniquely bears the apparent imprints of a crucified man. Unfortunately the cloth is incompatible with New

---

Joe Nickell's "Science *vs.* Shroud Science" originally appeared as "Scandals and Follies of the Holy Shroud" in the *Skeptical Inquirer* 25, no. 5 (September/October 2001).

Testament accounts of Jesus' burial. John's gospel (19:38–42, 20:5–7) specifically states that the body was "wound" with "linen clothes" and a large quantity of burial spices (myrrh and aloes). Still another cloth (called "the napkin") covered his face and head. In contrast, the Shroud of Turin represents a *single*, *draped* cloth (laid under and then over the "body") without any trace of the burial spices.

There were many earlier purported shrouds of Christ, which were typically about half the length of the Turin cloth.

One was the subject of a reported seventh-century dispute on the island of Iona between Christians and Jews, both of whom claimed it. As adjudicator, an Arab ruler placed the alleged relic in a fire from which it levitated, unscathed, and fell at the feet of the Christians—or so says a pious tale. In medieval Europe alone, there were "at least forty-three 'True Shrouds'" (Humber 1978, 78).

Scandal at Lirey

The cloth now known as the Shroud of Turin first appeared about 1355 at a little church in Lirey, in north central France. Its owner, a soldier of fortune named Geoffroy de Charney, claimed it as the authentic shroud of Christ, although he was never to explain how he acquired such a fabulous possession. According to a later bishop's report, written by Pierre D'Arcis to the Avignon pope, Clement VII, in 1389, the shroud was being used as part of a faith-healing scam:

> The case, Holy Father, stands thus. Some time since in this diocese of Troyes the dean of a certain collegiate church, to wit, that of Lirey, falsely and deceitfully, being consumed with the passion of avarice, and not from any motive of devotion but only of gain, procured for his church a certain cloth cunningly painted, upon which by a clever sleight of hand was depicted the twofold image of one man, that is to say, the back and the front, he falsely declaring and pretending that this was the actual shroud in which our Savior Jesus Christ was enfolded in the tomb, and upon which the whole likeness of the Savior had remained thus impressed together with the wounds which He bore. . . . And further to attract the multitude so that money might cunningly be wrung from them, pretended miracles were worked, certain men being hired to represent themselves as healed at the moment of the exhibition of the shroud.

D'Arcis continued, speaking of a predecessor who conducted the investigation and uncovered the forger: "Eventually, after diligent inquiry and examination, he discovered the fraud and how the said cloth had been cunningly painted, *the truth being attested by the artist who had painted it,* to wit, that it was a work of human skill and not miraculously wrought or bestowed" (emphasis added). Action had been taken and the cloth hidden away, but now, years later, it had resurfaced. D'Arcis (1389) spoke of "the grievous nature of the scandal, the contempt brought upon the Church and ecclesiastical jurisdiction, and the danger to souls."

As a consequence Clement ordered that, while the cloth could continue being exhibited (it had been displayed on a high platform flanked by torches), during the exhibition it must be loudly announced that "it is not the True Shroud of Our Lord, but a painting or picture made in the semblance or representation of the Shroud" (Humber 1978, 100). Thus the scandal at Lirey ended—for a time.

Further Misrepresentation

During the Hundred Years' War, Margaret de Charney, granddaughter of the Shroud's original owner, gained custody of the cloth, allegedly for safekeeping. But despite many subsequent entreaties she refused to return it, instead even taking it on tour in the areas of present-day France, Belgium, and Switzerland. When there were additional challenges to the Shroud's authenticity, Margaret could only produce documents officially labeling it a "representation."

In 1453, at Geneva, Margaret sold the cloth to Duke Louis I of Savoy. Some Shroud proponents like to say Margaret "gave" the cloth to Duke Louis, but it is only fair to point out that in return he "gave" Margaret the sum of two castles. In 1457, after years of broken promises to return the cloth to the canons of Lirey and later to compensate them for its loss, Margaret was excommunicated. She died in 1460.

The Savoys (who later comprised the Italian monarchy and owned the shroud until it was bequeathed to the Vatican in 1983) represented the shroud as genuine. They treated it as a "holy charm" having magical powers and enshrined it in an expanded church at their castle at Chambéry. There in 1532 a fire blazed through the chapel and before the cloth was rescued a blob of molten silver from the reliquary burned through its forty-eight folds. The alleged talisman was thus revealed unable even to protect itself. Eventually, in a shrewd political move—by a later duke who wished a more suitable capital—the cloth was transferred to Turin (in present-day Italy).

In 1898 the shroud was photographed for the first time, and the glass-plate negatives showed a more lifelike, quasi-positive image. Thus began the modern era of the shroud, with proponents asking how a mere medieval forger could have produced a perfect "photographic" negative before the development of photography. In fact the analogy with photographic images is misleading since the "positive" image shows a figure with white hair and beard, the opposite of what would be expected for a Palestinian Jew in his thirties.

Nevertheless, some shroud advocates suggested the image was produced by simple contact with bloody sweat or burial ointments. But that is disproved by a lack of wraparound distortions. Also, not all imaged areas would have been touched by a simple draped cloth, so some sort of *projection* was envisioned. One notion was "vaporography," body vapors supposedly interacting with spices on the cloth to yield a vapor "photo," but all experimentation produced was a blur (Nickell 1998, 81–84). Others began to opine that the image was "scorched" by a miraculous burst of radiant energy at the time of Jesus' resurrection. Yet no known radiation would produce such superficial images, and actual scorches on the cloth from the fire of 1532 exhibit strong reddish fluorescence, in contrast to the shroud images which do not fluoresce at all.

Secret Commission

In 1969 the Archbishop of Turin appointed a secret commission to examine the shroud. That fact was leaked, then denied, but (according to Wilcox 1977, 44) "At last the Turin authorities were forced to admit what they previously denied." The man who had exposed the secrecy accused the clerics of acting "like thieves in the night." More detailed studies—again clandestine—began in 1973.

The commission included internationally known forensic serologists who made heroic efforts to validate the "blood," but all of the microscopical, chemical, biological, and instrumental tests were negative. This was not surprising since the stains were suspiciously still red and artistically "picturelike." Experts discovered reddish granules that would not even dissolve in reagents that dissolve blood, and one investigator found traces of what appeared to be paint. An art expert concluded that the image had been produced by an artistic printing technique.

The commission's report was withheld until 1976 and then was largely suppressed, while a *rebuttal* report was freely made available. Thus began an approach that would be repeated over and over: distinguished experts would be asked to examine the cloth, then would be attacked when they obtained other than desired results.

Science versus "Shroud Science"

Further examinations were conducted in 1978 by the Shroud of Turin Research Project (STURP). STURP was a group of mostly religious believers whose leaders served on the Executive Council of the Holy Shroud guild, a Catholic organization that advocated the "cause" of the supposed relic. STURP members, like others calling themselves "sindonologists" (i.e., shroudologists), gave the impression that they started with the desired answer.

STURP pathologist Robert Bucklin—another Holy Shroud Guild executive councilman—stated he was willing to stake his reputation on the shroud's authenticity. He and other pro-shroud pathologists argued for the image's anatomical correctness, yet a footprint on the cloth is inconsistent with the position of the leg to which it is attached, the hair falls as for a standing rather than a recumbent figure, and the physique is so unnaturally elongated (similar to figures in Gothic art!) that one pro-shroud pathologist concluded Jesus must have suffered from Marfan's syndrome (Nickell 1989)!

STURP lacked experts in art and forensic chemistry—with one exception: famed microanalyst Walter C. McCrone. Examining thirty-two tape-lifted samples from the shroud, McCrone identified the "blood" as tempera paint containing red ocher and vermilion along with traces of rose madder—pigments used by medieval artists to depict blood. He also discovered that on the image—but not the background—were significant amounts of the red ocher pigment. He first thought this was applied as a dry powder but later concluded it was a component of dilute paint applied in the medieval *grisaille* (monochromatic) technique (McCrone 1996; cf. Nickell 1998). For his efforts McCrone was held to a secrecy agreement, while statements were made to the press that there was no evidence of artistry. He was, he says, "drummed out" of STURP.

STURP representatives paid a surprise visit to McCrone's lab to confiscate his samples, then gave them to two late additions to STURP, John Heller and Alan Adler, neither of whom was a forensic serologist or a pigment expert.

The pair soon proclaimed they had "identified the presence of blood." However, at the 1983 conference of the prestigious International Association for Identification, forensic analyst John F. Fischer explained how results similar to theirs could be obtained from tempera paint.

A more recent claim concerns reported evidence of human DNA in a shroud "blood" sample, although the Archbishop of Turin and the Vatican refused to authenticate the samples or accept any research carried out on them. University of Texas researcher Leoncio Garza-Valdez, in his *The DNA of God?* (1999, 41), claims it was possible "to clone the sample and amplify it," proving it was "ancient" blood "from a human being or high primate," while Ian Wilson's *The Blood and the Shroud* (1998, 91) asserted it was "human blood."

Actually the scientist at the DNA lab, Victor Tryon, told *Time* magazine that he could not say how old the DNA was or that it came from blood. As he explained, "Everyone who has ever touched the shroud or cried over the shroud has left a potential DNA signal there." Tryon resigned from the new shroud project due to what he disparaged as "zealotry in science" (Van Biema 1998, 61).

Pollen Fraud?

McCrone would later refute another bit of pro-shroud propaganda: the claim of a Swiss criminologist, Max Frei-Sulzer, that he had found certain pollen grains on the cloth that "could only have originated from plants that grew exclusively in Palestine at the time of Christ." Earlier Frei had also claimed to have discovered pollens on the cloth that were characteristic of Istanbul (formerly Constantinople) and the area of ancient Edessa—seeming to confirm a "theory" of the shroud's missing early history. Wilson (1979) conjectured that the shroud was the fourth-century Image of Edessa, a legendary "miraculous" imprint of Jesus' face made as a gift to King Abgar. Wilson's notion was that the shroud had been folded so that only the face showed and that it had thus been disguised for centuries. Actually, had the cloth been kept in a frame for such a long period there would have been an age-yellowed, rectangular area around the face. Nevertheless Frei's alleged pollen evidence gave new support to Wilson's ideas.

I say *alleged* evidence since Frei had credibility problems. Before his death in 1983 his reputation suffered when, representing himself as a handwriting expert, he pronounced the infamous "Hitler diaries" genuine; they were soon exposed as forgeries.

In the meantime an even more serious question had arisen about Frei's pollen evidence. Whereas he reported finding numerous types of pollen from Palestine and other areas, STURP's tape-lifted samples, taken at the same time, showed few pollen. Micropaleontologist Steven D. Schafersman was probably the first to publicly suggest Frei might be guilty of deception. He explained how unlikely it was, given the evidence of the shroud's exclusively European history, that thirty-three different Middle Eastern pollens could have reached the cloth, particularly only pollen from Palestine, Istanbul, and the Anatolian steppe. With such selectivity, Schafersman stated, "these would be miraculous winds indeed." In an article in *Skeptical Inquirer* Schafersman (1982) called for an investigation of Frei's work.

When Frei's tape samples became available after his death, McCrone was asked to authenticate them. This he was readily able to do, he told me, "since it was easy to find red ocher on linen fibers much the same as I had seen them on my samples." But there were few pollen other than on a single tape which bore "dozens" in one small area. This indicated that the tape had subsequently been "contaminated," probably deliberately, McCrone concluded, by having been pulled back and the pollen surreptitiously introduced.

McCrone added (1993):

> One further point with respect to Max which I haven't mentioned any-where, anytime to anybody is based on a statement made by his coun-terpart in Basel as head of the Police Crime Laboratory there that Max had been several times found guilty and was censured by the Police hier-archy in Switzerland for, shall we say, overenthusiastic interpretation of his evidence. His Basel counterpart had been on the investigating com-mittee and expressed surprise in a letter to me that Max was able to con-tinue in his position as Head of the Police Crime Lab in Zurich.

C-14 Falsehoods

The pollen "evidence" became especially important to believers following the devastating results of radiocarbon dating tests in 1988. Three laboratories (at Oxford, Zurich, and the University of Arizona) used accelerator mass spec-trometry (AMS) to date samples of the linen. The results were in close agree-ment and were given added credibility by the use of control samples of known dates. The resulting age span was circa A.D. 1260–1390—consistent with the time of the reported forger's confession.

Shroud enthusiasts were devastated, but they soon rallied, beginning a campaign to discredit the radiocarbon findings. Someone put out a false story that the AMS tests were done on one of the patches from the 1532 fire, thus supposedly yielding a late date. A Russian scientist, Dmitrii Kuznetsov, claimed to have established experimentally that heat from a fire (like that of 1532) could alter the radiocarbon date. But others could not replicate his alleged results and it turned out that his physics calculations had been plagiarized—complete with an error (Wilson 1998, 219–23). (Kuznetsov was also exposed in *Skeptical Inquirer* for bogus research in a study criticizing evolution [Larhammar 1995].)

A more persistent challenge to the radiocarbon testing was hurled by Garza-Valdez (1999). He claimed to have obtained a swatch of the "miraculous cloth" that bore a microbial coating, contamination that could have altered the radiocarbon date. However, that notion was effectively debunked by physicist Thomas J. Pickett (1996). He performed a simple calculation which showed that, for the shroud to have been altered by thirteen centuries (i.e., from Jesus' first-century death to the radiocarbon date of 1325±65 years), there would have to be twice as much contamination, by weight, as the cloth itself!

Shroud of Rorschach

Following the suspicious pollen evidence were claims that plant images had been identified on the cloth. These were allegedly discerned from "smudgy"-appearing areas in shroud photos that were subsequently enhanced. The work was done by a retired geriatric psychiatrist, Alan Whanger, and his wife, Mary, former missionaries who have taken up image analysis as a hobby. They were later assisted by an Israeli botanist who looked at their photos of "flower" images (many of them "wilted" and otherwise distorted) and exclaimed, "Those are the flowers of Jerusalem!" Apparently no one has thought to see if some might match the flowers of France or Italy or even to try to prove that the images are indeed floral (given the relative scarcity of pollen grains on the cloth).

The visualized "flower and plant images" join other perceived shapes seen—Rorschach-like—in the shroud's mottled image and off-image areas. These include "Roman coins" over the eyes, head and arm "phylacteries" (small Jewish prayer boxes), an "amulet," and such crucifixion-associated items (cf.

John, chap. 19) as "a large nail," a "hammer," "sponge on a reed," "Roman thrusting spear," "pliers," "two scourges," "two brush brooms," "two small nails," "large spoon or trowel in a box," "a loose coil of rope," a "cloak" with "belt," a "tunic," a pair of "sandals," and other hilarious imaginings including "Roman dice"—all discovered by the Whangers (1998) and their botanist friend.

They and others have also reported finding ancient Latin and Greek words, such as "Jesus" and "Nazareth." Even Ian Wilson (1998, 242) felt compelled to state: "While there can be absolutely no doubting the sincerity of those who make these claims, the great danger of such arguments is that researchers may 'see' merely what their minds trick them into thinking is there."

Conclusion

We see that "Shroud science"—like "creation science" and other pseudo-sciences in the service of dogma—begins with the desired answer and works backward to the evidence. Although they are bereft of any viable hypothesis for the image formation, sindonologists are quick to dismiss the profound, cor-roborative evidence for artistry. Instead, they suggest that the "mystery" of the shroud implies a miracle, but of course that is merely an example of the logi-cal fallacy called arguing from ignorance.

Worse, some have engaged in pseudoscience and even, apparently, out-right scientific fraud, while others have shamefully mistreated the honest sci-entists who reported unpopular findings. We should again recall the words of Canon Ulysse Chevalier, the Catholic scholar who brought to light the docu-mentary evidence of the shroud's medieval origin. As he lamented, "The his-tory of the shroud constitutes a protracted violation of the two virtues so often commended by our holy books: justice and truth."

References

Antonacci, Mark. 2000. *The Resurrection of the Shroud: New Scientific, Medical and Ar-cheological Evidence.* New York: M. Evans and Co.

Binga, Timothy. 2001. "Report in Progress from the Director of the Center for Inquiry Libraries." June 19.

D'Arcis, Pierre. 1389. Memorandum to the Avignon Pope, Clement VII; translated from Latin by Rev. Herbert Thurston, reprinted in Wilson 1979, 266–72.

Garza-Valdez, Leoncio A. 1993. *Biogenic Varnish and the Shroud of Turin.* Cited in

Garza-Valdez 1999, 37.

———. 1999. *The DNA of God?* New York: Doubleday.

Gove, Harry E. 1996. *Relic, Icon, or Hoax? Carbon Dating the Turin Shroud.* Philadelphia: Institute of Physics Publishing.

Humber, Thomas. 1978. *The Sacred Shroud.* New York: Pocket Books.

Larhammar, Dan. 1995. "Severe Flaws in Scientific Study Criticizing Evolution." *Skeptical Inquirer* 19, no. 2 (March/April): 30–31.

McCrone, Walter C. 1993. Letters to Joe Nickell, June 11 and 30.

———. 1996. *Judgement Day for the Turin "Shroud."* Chicago: Microscope Publications.

Nickell, Joe. 1998. *Inquest on the Shroud of Turin: Latest Scientific Findings.* Amherst, N.Y.: Prometheus Books. (Except as otherwise noted, information is taken from this source.)

———. 1989. "Unshrouding a Mystery: Science, Pseudoscience, and the Cloth of Turin." *Skeptical Inquirer* 13, no. 3 (spring): 296–99.

Pickett, Thomas J. 1996. "Can Contamination Save the Shroud of Turin?" *Skeptical Briefs* (June): 3.

Whanger, Mary, and Alan Whanger. 1998. *The Shroud of Turin: An Adventure of Discovery.* Franklin, Tenn.: Providence House.

Wilcox, Robert K. 1977. *Shroud.* New York: Macmillan.

Wilson, Ian. 1979. *The Shroud of Turin,* rev. ed. Garden City, N.Y.: Image Books.

———. 1998. *The Blood and the Shroud.* New York: Free Press.

VI

Scientific Explanations
of Religious Belief

30
WHY DO PEOPLE
BELIEVE OR DISBELIEVE?

PAUL KURTZ

Scientists attempt to account for various forms of human behavior: economic, political, sociobiological, psychological. Why not deal with religious behavior in naturalistic terms? If we can talk about political science, or economic science, can we also talk about the science of religion?

The answers to this question, I submit, are in the affirmative, at least in principle. Indeed, a great deal of scientific energy has already been expended to account for behavior that is described as "religious." Historians attempt to describe the historical past of religious institutions; sociologists seek to explain their social structures and functions; anthropologists deal with religions in primitive cultures; psychologists of religion seek to account for "religious experience," the role of prayer and ritual, in human behavior, and so forth. There are any number of scholarly and scientific fields that deal with the historical and archaeological contexts in which the sacred texts were written or engage in philological and comparative analyses. The premise of all these studies is that we should treat the varieties of religious behavior as we would any other forms of human behavior; that is, approach them objectively and dispassionately,

Paul Kurtz's "The Science of Religion: Why Do People Believe or Disbelieve?" is adapted from his article "Why Do People Believe or Disbelieve?" which appeared in *Free Inquiry* 19, no. 3 (summer 1999).

attempting to understand what is going on without an a priori evaluative bias. Moreover, so-called supernatural claims that have some empirical reference can be examined as "paranatural" phenomena.

I wish in this chapter to focus on two basic questions: Why do people believe in religious doctrines, i.e., why do they accept the tenets of a religion and participate in its practices and rituals? Conversely, we may ask, Why do some people *dis*believe in the tenets of religion and/or reject its practices? Religion is one of the most pervasive and enduring characteristics of human culture. Predictions by scientists and rationalists at the end of the eighteenth and nineteenth centuries that religion would eventually disappear surely were premature because by the end of the twentieth century it remained as strong as ever.

Religion's Power

Three dramatic religious events illustrate the perennial power of religious faith in human culture. The first is the huge annual assembly of Islamic pilgrims drawn to Mecca every year. Photographs of an estimated three million devotees who were in Mecca in recent pilgrimages show that they have come from all walks of life and from all classes. The second impressive annual event is the millions of Hindus in India who congregate at the Ganges River in accordance with ancient religious rituals. At a recent event, an estimated ten million people appeared at the Kumbha Mela festival in the small city of Hardiwar for prayer and purification. And the third is the re-exhibition at a cathedral in Turin of the shroud that Jesus was allegedly wrapped in and buried. A huge throng of visitors have come from all over the world to view the Shroud of Turin.

Skeptical doubts can surely be raised about the claim that a pilgrimage to Mecca will guarantee Muslim believers entrance to heaven and/or that bathing in the Ganges River will bestow special spiritual benefits. These are sheer acts of faith drawing upon ancient traditions that scientific skeptics would maintain have little basis in empirical fact. There is no evidence that the performance of ritualistic acts of spiritual contrition, either by visiting the Kaaba in Mecca and encircling it three times, or by bathing in the water of the Ganges, will achieve a blessed state of Paradise for Muslims or Atman for Hindus. Remarks to the devout disciples of these two ancient religions that the recommended rites are contradictory or have no basis in fact generally fall on deaf ears.

Similarly for the Shroud of Turin, which, according to the best available scientific evidence, was a forgery made in Lirey, France, in the fourteenth century.[1] Interestingly, it was condemned as such at that time by the bishop in the area, for it was used to deceive thousands of pilgrims seeking cures for their illnesses. Walter McCrone, the noted microscopist, has shown that the red color on the Shroud was not human blood, but red ochre and vermilion tempera paint. Joe Nickell has even demonstrated that it is possible to produce a similar image on cloth by a rubbing technique, using the vermilion and ochre pigments that were available at that time in France. Moreover, portions of the Shroud were carbon-14 dated by three independent laboratories, all of whom reported that it was not 1,900 years old, but probably fabricated approximately 700 years ago. These reports were published in the scientific literature and received widespread attention in the press; and skeptical scientists applauded the forensic evidence, which clearly stated that the image on the Shroud was not due to a miracle, but could be given a naturalistic causal explanation.

Yet, much to the surprise of skeptics, who thought that they had decisively refuted the proponents of the faith, the Shroud industry has returned with full force and vigor and is again proclaiming that skeptics were in error. Believers maintain that there were alleged flaws in the carbon-14 process—all rationalizations in the view of skeptics—and that the Shroud was the burial garment of Jesus Christ.

The History of Religion

Why do people believe in the above religious claims? Is it because they have not been exposed to criticisms? Most of the classical religious beliefs emerged in a prescientific era before the application of the methods of science. Unfortunately, the origins of the venerated ancient religions are often buried by the sands of historical time—though biblical critics have endeavored to reconstruct the foundations of these religions by using the best scholarly and scientific methods of inquiry. It is often difficult to engage in impartial scholarly or scientific inquiry into the origins of religious doctrines, particularly when those critically examining the foundations of the revered truths are often placed in jeopardy by their societies. Biblical criticism in the Western world has only relatively recently been freed from prohibiting censorship and/or the power of institutional sanctions brought to bear on freethinkers. Koranic criticism is virtually absent in Islamic lands, or if it is done it is only with great fear of retribution; for questioning the divine authority of Muhammad is considered by the Koran itself to be a form of blasphemy punishable by a *fatwa*.

The ancient religions of prophecies and revelations—Judaism, Christianity, and Islam—all claim that God intervened at one time in history, spoke to Moses and the prophets, resurrected Jesus, or communicated through Gabriel to Muhammad. Skeptics maintain that these paranatural claims have never been adequately corroborated by reliable independent eyewitnesses. The so-called sacred books no doubt incorporate the best theological and metaphysical yearnings of ancient nomadic and agricultural societies, and they often express eloquent moral insights by the people of that time; yet they hardly can withstand the sustained critical examination by objective inquirers. The narratives of alleged supernatural intervention that appear in the Bible and the Koran were at first transmitted by oral traditions after the alleged facts occurred. They were written down by second- or thirdhand sources, many years and even decades later. They most likely weave into their parables dramatic renditions bordering on fiction, and were written by passionate propagandists for new faiths. These sacred books promise believers another world beyond this vale of tears. Their messages of salvation were attractive to countless generations of poor and struggling souls endeavoring to overcome the blows of existential reality. Believers ever since have accepted them as gospel truth; after centuries they became deeply ingrained in the entire fabric of society. Indeed, the great monotheistic religions were eventually intertwined with the dominant political, military, and economic institutions and were enforced by both priestly and secular authorities.

The religion of the ancient Jews, allegedly inspired by Moses and the Old Testament prophets, came to express the ideological yearnings of the Hebrew nation. Christianity was eventually declared to be the state religion by Constantine. Islam, from its inception, was reinforced by the sword of Muhammad. All these faiths, though shrouded in mystery, claim divine sanctification. There are certain common features which each of these religions manifests—historic claims of revelation by charismatic prophets promising eternal salvation; sacred books detailing their miraculous prophecies, prescribing rituals, prayers, and rites of passage; a priestly class which seeks to enforce religious law; great temples, cathedrals, and mosques where the Lord is present in the mysteries of the sacraments. These ancient religions have persisted in part because they have ostracized or condemned heretics and disbelievers. They have gained adherents over time by policies of selective breeding: marriage could only be by members of the same clan or tribe or church, disowning those who married outside of the faith. They sought to inculcate and transmit the tenets of the faith to the young, so as to ensure the continuity of the tradition. The entire artistic, moral, philosophical, economic, social, and legal structures of ancient societies were rooted in religious institutions.

Many liberal theists would accept the above critique of the historic religions by the "higher criticism," especially since the German theologian Rudolf Bultmann attempted to demythologize the New Testament. Yet they maintain that the alleged historical events are to be read symbolically or metaphorically and if they are accepted it is because they give meaning and purpose to life.

Interestingly, we now have data from recent religious sects that emerged in the nineteenth century and are not shrouded in historical mystery. And we are close enough to the events to lay bare the factors at work: the historical records of persuasion and conversion on the part of the founders of these new religions, and the willing acceptance of the faith by receptive believers. Thus we may examine the origins of Mormonism, Seventh-day Adventism, Christian Science, or the Jehovah's Witness movement to discern if there are similar psycho-bio-sociological patterns at work. Invariably it is difficult to certify their authenticity once the claims to divine revelation are examined by careful historical investigators.

In many new religions the historical records are abundant. In all of these religions, critics have pointed out the role of deception or self-deception, such as Joseph Smith's writing of the Book of Mormon and his accounts of the golden plates delivered by the angel Moroni, which were subsequently lost by him. Similarly for the claims of plagiarism made against Mary Ellen White, founder of the Seventh-day Adventist Church, or the questionable claims of miraculous health cures by Mary Baker Eddy and other Christian Scientist practitioners. Similarly for the origins of Jehovah's Witnesses.

Closer still, twentieth-century skeptics have been able to witness first-hand the spinning out of New Age paranormal religions. A good illustration of this is the power of suggestion exercised by psychics and mediums, often through the use of deception or self-deception, and the receptiveness of so many believers, all too willing to accept claims of supernormal powers by abandoning rigorous standards of corroboration. These processes are even found among sophisticated scientists as well as ordinary folks, who are specialists in other fields, but perhaps not in the art of deception. An entire industry claiming to prove another reality transcending this world is flourishing: belief in reincarnation is based on "past-life regressions," and near-death experiences are often appealed to in order to reinforce belief in the separable existence and immortality of the human soul.

The spawning of the space-age religions in the latter half of the twentieth century is especially instructive for the psychobiology of belief. Scientology was invented by Ron Hubbard, who began as a writer of science fiction but then went on to consciously create a new religion. Dianetics and all that it proposes are questionable on empirical grounds; yet countless thousands of people, including famous celebrities, have been persuaded to accept its tenets. UFO mythology is especially fascinating. Space Age prophets have emerged, rivaling the classical religious prophets, and likewise claiming deliverance to another realm. This is the age of the great human adventure of space exploration, and so people are conscious of other planets in our solar system and other galaxies far beyond. It is also a time in which astronomy has made great strides and telescopes have enabled humans to extend the reach of observations. It is also an age in which science fiction has soared far beyond verifiable hypotheses and in which the speculative creative imagination is unbounded. Beginning with the premise that it is possible, indeed probable, that life, even intelligent life, exists elsewhere in the universe, there is a leap of faith to the conviction that the planet Earth has been visited by extraterrestrial aliens, that some Earthlings have been abducted, and that intergalactic biogenetic breeding has occurred. Thus the possible has been converted into the actual and fiction transposed into reality. Extraterrestrial visitations from on high have the similar contours of alleged early visitations by divine beings and their revelations on Mount Sinai or in the caves of Hijra outside of Mecca, or on the road to Damascus, or by the Olympian gods of Greek mythology.

Thus the question is raised anew, How do we explain the willingness of so many people—no doubt a majority of humankind—to outstrip the evidence and to weave out fantasies in which their deepest psychological longings are expressed and their national mythologies fulfilled? How explain the willingness to believe even the most bizarre tales?

I have had close contact over the years with a wide range of latter-day religious gurus and mystics—from Reverend Moon to Ernest Angley and Peter Popoff—and paranormal psychics and seers—from Uri Geller to Jeane Dixon, Ramtha, and Raël. Skeptics have been challenged to account for the apparent extraordinary feats of these proponents. After detailed investigation their weird claims have been refuted; yet in spite of this, otherwise sensible people have persisted in beliefs that are patently false. Indeed, there seems to be a bizarre kind of logic at work: belief systems for which there is entirely scanty evidence or no evidence, or indeed abundant evidence to the contrary are fervently accepted; indeed, people will devote their entire lives to a groundless creed. This has been heralded in the past as *faith* in things unseen or things hoped for. The will to believe in spite of negative evidence has been acclaimed as morally praiseworthy. David Hume thought it a "miracle" that people who believe in miracles are willing to subvert all of the evidence of the senses and the processes of rationality in order to accept their beliefs.

Explaining Belief

There are *at least* two possible explanations that I wish to focus on. There are no doubt others, such as the need for identity, the quest for community, the role of indoctrination, the power of tradition, ethnicity, and so on, that I will not discuss here.

In answer to the question, Why do people believe? the first explanation is that believers have not been exposed to the factual critiques of their faith. These critiques apply to the cognitive basis of their belief. There are alternative naturalistic explanations of the alleged phenomena, cognitivists maintain, and if criticisms of the claims were made available to them, they would abandon their irrational beliefs. This is no doubt true of some people who are committed to inquiry, but not of all, for processes of rationalization intervene to rescue the faith.

Accordingly, a second explanation for this is that noncognitive tendencies and impulses are at work, tempting believers to accept the "unbelievable." This disposition to believe in spite of insufficient or contrary evidence has deep roots in our biological and social nature.

In the first instance, cognition performs a powerful role in human life, liberating us from false ideas. In the form of common sense, it is essential, at least up to a point, if we are to live and function in the real world. Ordinary men and women constantly appeal to practical reason to refute unwarranted beliefs. They are forced to maintain some cognitive touch with reality if they are to survive in the natural and social environment. Human beings are capable of some rational thought, and this is the most effective capacity that they have for coping with obstacles that are encountered. Critical thinking is the preeminent instrument of human action; it is the most effective means that we have to fulfill our purposes and solve the problems of living. Cognition is the most powerful method for making sense of the world in which we live. From it philosophy and science have emerged, contributing to our understanding of nature and ourselves.

We all know that we need to use practical reason to deal with empirical questions, such as: "Is it raining outside?" or "How do I cope with my toothache?" And we also apply such methods within the sciences, to deal with issues such as the following: "The dinosaurs were most likely extinguished by an asteroid impact some sixty-five million years ago," or, "We are unable to cure people by therapeutic touch." Each of these beliefs may be tested by the experimental evidence or by theories accepted as probable or improbable on the basis of these considerations. In addition, an open-minded inquirer may be led to accept or reject any number of propositions, which he or she previously asserted, such as, "There is no evidence that a great flood engulfed the entire globe as related in the Bible."

There is a class of *overbeliefs*, however, for which no amount of evidence seems to suffice, at least for some people. These generally may be classified as "transcendental beliefs." It is here that faith or the will to believe intervenes. By the "transcendental" I mean that which is over and beyond normal observations or rational coherence, and is enhanced by mystery and magic. This surely is what the great mystics have referred to as the "ineffable" depths of Being. Scientific inquiry is naturalistic; that is, it attempts to uncover the natural causes at work. Granted that these are often hidden causes, unseen by unaided observation, such as microbes or atoms; yet such causes can be confirmed by some measure of verification; they fit into a conceptual framework; and their explanatory value can be corroborated by a community of independent inquirers. Transcendental explanations are, by definition, nonnatural; they cannot be confirmed experimentally; they cannot be corroborated objectively.

We may ask, "Why do many people accept unverified occult explanations when they are clothed in religious, supernatural, or paranormal guise?" The answer, I think, in part at least, *is because such accounts arouse awe and entice the passionate imagination.* I have in my earlier writings labeled this "the transcendental temptation,"[2] the temptation to believe in things unseen, because they satisfy needs and desires. The transcendental temptation has various dimensions. It was resorted to by primitive men and women, unable to cope with the intractable in nature, unmitigated disasters, unbearable pain or sorrow. It is drawn upon by humans in order to assuage the dread of death—by postulating another dimension to existence, the hope for an afterlife in which the evils and injustices of this world are overcome. The lure of the transcendental appeals to the frail and forlorn. There may not be any evidence for it; but powerful emotive and intellectual desire to submit to it can provide a source of comfort and consolation. To believe that we will meet in another life those whom we have loved in this life can be immensely satisfying, or at least it can provide some saving grace. It may enable a person to get through the grievous losses that he or she suffers in this life. If I can't be with those I cherish today, I can at least do so in my dreams and fantasies, and if I submit to and propitiate the unseen powers that govern the universe this will miraculously right the wrongs that I have endured in this vale of tears. Thus the lure of the transcendental is tempting because it enables human beings to survive the often cruel trials and tribulations that are our constant companion in this life, and it enables us to endure in anticipation of the next. It is the mystery and magic of religion, its incantations and rituals, that fan the passions of overbelief, and nourish illusion and unreality. There is a real and dangerous world out there that primitive and modern humans need to cope with—wild animals and marauding tribes, droughts and famine, lightning and forest fires, calamities and deprivation, accidents and contingencies. Surely, there is pleasure and satisfaction, achievement and realization in life, but also tragedy and failure, defeat and bitterness. Our world is a complex tapestry of joy and suffering. The transcendental temptation thus can provide a powerful palliative enabling humans to cope with the unbearable, overcome mortality and finitude; and it does so by creating fanciful systems of religious overbelief in which priests and prophets propitiate the unseen sources of power and thus shield us from the vicissitudes of fortune. Humans tend to corrupt their visions of reality, according to psychologist John Schumaker, in order to survive in a world that they cannot fully comprehend.[3]

It is only in recent human history that the species has gradually been able to overcome mythological explanations. Philosophy and metaphysics emerged, attempting to account for the world of change and flux in terms of rational explanations; modern science succeeded where pure speculation failed, by using powerful cognitive methods of experimental verification and mathematical inference. What had been shrouded in mystery was now explicable in terms of natural causes. Diseases did not have satanic origins, but natural explanations and cures. The weather could be interpreted, not as a product of divine wrath or favor, but in meteorological terms. Nature could be accounted for by locating the natural causes of phenomena. Astrology's heavenly omens and signs were replaced by the regularities discernible by physics and astronomy. Science abandoned occult for material causes. It is the foe of magical thinking, and it is able to proceed by refusing to submit to transcendental deception, at least in dealing with the empirical world. Thus there has been a continuous retreat of magical thinking under the onslaught of cognitive inquiry. The same methods of inquiry used so successfully in the natural sciences were extended to biology and the social sciences. Science thus continues to make progress by using rigorous methods of naturalistic inquiry.

Yet there still remained a residue of unanswered questions, and it is here in the swamp of the unknowable that the transcendental temptation persists. This beguiling temptation reaches beyond the natural world by sheer force of habit and passion, and it resists all efforts to contain it. Rather than suspend judgments about those questions for which there is no evidence either way, it leaps in to fill the void and comfort the aching soul. It is the most frequent salve used to calm existential fear and trembling. Why is this so? Because I think that the *temptation* has its roots in a *tendency*, and this in a *disposition*. In other words, there may be within the human species a *genetic* component, which is stronger than temptation and weaker than instinct. The hypothesis that I wish to offer is that the belief in the efficacy of prayer and the submission to divine power persists because it has had some survival value in the infancy of the race; powerful psycho-socio-biological factors are thus at work, predisposing humans to submit to the temptation.

The cognitive explanation for its persistence is that there is cognitive dissonance or misinformation that is the root cause for the fixation on the transcendental and that this can be overcome by rational inquiry. Socrates thought that faith persisted only because of ignorance, and that knowledge would disabuse us of religious myths. This surely continues to play a powerful role in regard to the content of our beliefs. Yet I submit that there is another factor present which explains the persistence of religiosity, and this is an evolutionary explanation; that is, belief in the transcendental had adaptive value, and those tribes or clans which believed in unseen myths and forces to whom they propitiated by ritual and prayer had a tendency to survive and to pass on this genetic predisposition to their offspring. Thus religiosity is a "heritable" factor within the naked human ape.[4]

What are some of the data in support of a transcendental predisposition? There are the University of Minnesota studies of identical twins,[5] which showed that a significant number of infants who were separated at birth and reared apart under different environmental conditions nonetheless exhibited similar tastes and preferences, and in this case exhibited a tendency to be religious. This predisposition is not necessarily deterministic in a strict sense, and it is absent in a number of cases. The heritable factor is estimated to be 50 percent. E. O. Wilson also maintains that there is some biological basis for religiosity; though one cannot locate this in a specific gene, there are a multiplicity of genetic factors and epigenetic rules.[6] He argues that theological overbeliefs offer consolation in the face of adversity, and that these religious overbeliefs—whether true or false—provide a functional means of adaptation. Those tribes or clans which possessed a safety net of such beliefs-practices may have been better able to cope with the fear of death, and they were also able to pass along to future generations the tendency to be religious. This proclivity may have had some survival value and thus it was transmitted to future generations. E. O. Wilson claims that "there is a hereditary selective advantage to members in a powerful group united by devout belief and purpose. . . . Much if not all religious behavior could have arisen from evolution by natural selection."[7]

There is a growing body of scientific research which supports this sociobiological explanation: this includes two components: (a) *psychobiological,* which has some genetic basis, and (b) *sociological,* which has roots in cultural memes and habits. This would involve a coeval gene-meme hypothesis. Evolution is a function of both our *genes* on the one hand and *memes* transmitted by culture and inculcated in the young on the other.[8] Thus, both hereditary and environmental factors have an influence on the behavior of individuals. Though there may be a predisposition toward belief in the transcendental, *how* it is expressed and the content of the beliefs depends on the culture.

The Reasons for Disbelief

We need also to ask, Why do some humans *dis*believe?—for there is a minority of people who remain unbelievers, agnostics, or atheists.[9] There are a number of important research projects that I think should be undertaken. To ascertain if there is a genetic tendency—or lack of it—we should study the family trees of both believers and unbelievers. Much the same as we can trace the physical characteristics, such as eye or hair color, short or tall stature, and even genetic diseases in some family stocks, so we should be able to trace the religiosity factor, especially in twins and/or siblings who are reared apart. If we can measure musical talent (MQ) or intelligence (IQ), then perhaps we can also measure the religious quotient (RQ). Similarly, we need to trace the family trees of unbelievers and ask, Is the genetic factor absent and if so to what extent and why?

I have met a great number of unbelievers over the years who tell me that they have been atheists for as long as they could remember, that they never could accept the dominant religious creed, even though many were indoctrinated into it from the earliest. Clearly, we need to go beyond anecdotal autobiographical accounts to systematic studies of how and why people become disbelievers.

I have conducted polls of atheists and secular humanists at numerous talks that I have delivered, and prima facie these seem to support the hypothesis for a significant number of unbelievers, but there needs to be more extensive polling. There are most likely a plurality of explanations. Many atheists have related that their unbelief was a result of a slow cognitive process of critical reflection. Bruce Hunsberger and Bob Altemeyer, in an important study,[10] have attempted to outline the processes of conversion and deconversion in students that they studied in universities in Canada. Edward Babinsky[11] has published autobiographical accounts of why people abandoned their religiosity. We need to study the processes of deconversion: Why do people who were religiously indoctrinated reject their beliefs, how rapidly did they do so, and for what reasons or causes? Conversely, what processes are involved in moving from a state of unbelief to religious conviction? No doubt there are many factors at work; we need to sort them out. Hunsberger and Altemeyer have suggested in their study of students that the process of deconversion was predominantly a slow, cognitive process; and that of conversion a rather rapid emotional transformation.

We need to examine the sociocultural contexts in which religious ideas appear and disappear. We have an excellent data pool today in Russia and eastern Europe where atheism was the official doctrine of the state. Here enormous efforts were expended for fifty to seventy-five years to pursue political policies of indoctrination and propaganda, designed to discourage religious belief and encourage atheism. We may ask, What has happened in these countries since the collapse of communism? Is the past political-social influence of atheism enduring, leaving a permanent residue, or is it dissipating?

Similarly, many western European countries have seen a rather rapid decline in traditional religion in the post–World War II period, especially under the influence of secular liberalism and humanism. For example, in the Netherlands before the war approximately half of the population identified with Roman Catholicism and half with Protestantism, with a small percentage of Jews and other minorities. This has changed since World War II where there is now a higher percentage of humanists then either Protestants or Catholics. Similar processes have been observed in Norway, England, France, and elsewhere.

A recent poll conducted by sociologists at the City University of New York Graduate Center in 2001 shows that 14.1 percent of the adult American population has no religious identification, an increase from 8 percent a decade earlier, which may suggest an analogous secular trend developing in the United States. Still, only 6 to 8 percent of the American population may be classified as unbelievers.[12] Can we give an account of why this is so and why American society seems to be anomalous, at least in comparison with western Europe? Inasmuch as a majority of American scientists, according to a recent poll, are classified as unbelievers, we may ask, Why does this happen? Are there cognitive factors primarily at work? Or are disbelievers aberrant—lacking the genetic disposition? Or on the contrary, do they represent an advanced form of the evolution of the species?[13]

A key factor in the growth of religion or atheism undoubtedly is a function of the sociocultural influences that prevail, and perhaps the American situation is rather unique, especially in comparison with many European democracies, which seem to be postreligious. Another consideration to bear in mind is that historically the orthodox religions have sought to punish heresy or blasphemy as high crimes. Infidels have often been excommunicated or burned at the stake. It is only in recent times that democratic societies have recognized, let alone permitted or encouraged, religious dissenters to flourish.[14] One might ask, If the condition of tolerance, indeed encouragement, were to prevail—as in much of Europe—to what extent would religious beliefs wane or be altered? How can this be developed? What are the environmental conditions under which atheism would be induced? What kinds of educational curricula would most likely stimulate unbelief?

A further issue that can be raised concerns the difference between the *content* of the core beliefs and practices of a religion and the *function* of the beliefs and practices. The content may change over time, and there may be an erosion of traditional beliefs and their modification due to cognitive criticisms; but alternative creeds-practices may emerge, satisfying similar psycho-biological-sociological needs and functions. In this regard, I reiterate, we are not dealing with the *kind* of religion that persists or the status of its *truth claims*—which may be irrelevant for many believers—but with the *power* of religious symbols and institutions to provide structure and order, and to give purpose in an otherwise meaningless and perhaps terrifying universe.

If science confirms the hypothesis that there *are* deep sociobiological forces responsible, at least in part, for religiosity in the species, then we need to ask, What can naturalists do about it, if anything? Cognitivists will say that we still should constantly strive to engage in criticism of patently fake and destructive doctrines. At the very least this will help to restrain and temper religious fanaticism, protect the rights of unbelievers, and perhaps develop an ethic of tolerance. If religiosity will most likely be with us in one form or another in the foreseeable future, can secular and naturalistic substitutes or moral equivalents be developed for the passionate longing for meaning? Can they provide sufficient balm to soothe existential *weltschmerz*? Can new symbols to inspire meaning and hope be devised? Can the courage to be and to become emerge? In other words, can naturalistic humanism offer a message as potent as theistic mythology? These are the kinds of questions that science may help us to solve. But they are predicated on our understanding how and why people believe or disbelieve in a religion.

Notes

1. See Joe Nickell, *Inquest on the Shroud of Turin*, rev. ed. (Amherst, N.Y.: Prometheus Books, 1998); Walter McCrone, *Judgment Day for the Shroud of Turin* (Amherst, N.Y.: Prometheus Books, 1999). Invariably, it is difficult to certify their authenticity once the claims to divine revelation are examined by careful historical investigators.

2. Paul Kurtz, *The Transcendental Temptation: A Critique of Religion and the Paranormal* (Amherst, N.Y.: Prometheus Books, 1986).

3. John F. Schumaker, *The Corruption of Reality: A Unified Theory of Religion, Hypnosis and Psychopathology* (Amherst, N.Y.: Prometheus Books, 1995).

4. If it is the case that there is a genetic predisposition for religiosity, then we need an operational criterion of it. I would define (theistic) "religiosity" behavioristically: the expression of piety, the veneration of the mysterious beyond ordinary experience, the cherishing of overbeliefs about the transcendental, symbolic acts of submission to a divine figure(s) in expectation of receiving salvation, the engaging in propitiatory prayer and ritual.

5. N. G. Waller et al., "Genetics and Environmental Influences on Religious Interests, Attitudes, and Values: A Study of Twins Reared Apart and Together," *Psychological Science* (1990): 138–42; Thomas J. Bouchard Jr. et al., "Sources of Human Psychological Differences: The Minnesota Study of Twins Reared Apart," *Science* 250, no. 4978 (October 12, 1990): 223–28.

6. E. O. Wilson, *Consilience: The Unity of Knowledge* (New York: Alfred A. Knopf, 1998). See also John C. Avise, *The Genetic Gods: Evolution and Belief in Human Affairs* (Cambridge, Mass.: Harvard University Press, 1998).

7. E. O. Wilson, *Consilience*, p. 258.

8. Richard Dawkins has postulated "memes" as an explanation of the persistence of beliefs. Memes refer to culturally conditioned imitative behavior which sets patterns of images and beliefs in people, often from the earliest stages of childhood development. This has been supported by the work of Susan Blackmore.

9. The readers of *Free Inquiry* and the *Skeptical Inquirer* magazines provide a large pool of unbelievers, a good source for research. A poll of *Free Inquiry* readers indicates that 91 percent are either atheists, agnostics, or secular humanists, and of *Skeptical Inquirer*, 77 percent are atheists, agnostics, or secular humanists.

10. Bob Altemeyer and Bruce Hunsberger, *Amazing Conversions: Why Some Turn to Faith and Others Abandon Religion* (Amherst, N.Y.: Prometheus Books, 1997).

11. Edward T. Babinsky, *Leaving the Fold: Testimonies of Former Fundamentalists* (Amherst, N.Y.: Prometheus Books, 1994).

12. "Religious Belief in America: A New Poll," *Free Inquiry* 16, no. 3 (summer 1996): 34–40.

13. If one were to conclude that there was a heritability factor, and if one believed that atheism should be encouraged in the population, then one might wish to encourage atheists to marry atheists and to bring up the children as atheists, so as to increase the number of atheist offspring. Often a minority religion grows not so much by conversion but by outbreeding other sects.

14. See the book by David Berman, *A History of Atheism in Britain: From Hobbes to Russell* (New York: Routledge, 1990).

31
SUPERNATURAL POWER AND CULTURAL EVOLUTION

ANTHONY LAYNG

All human cultures appear to include faith in supernatural power, and it seems that this tradition has played an active and essential role in influencing how our ancestors perceived and adapted to their environment. This relationship between human belief and adaptive behavior may have played a critical role in shaping cultural evolution.

The history of any society can demonstrate how sacred beliefs change over time, how they are created or borrowed, consciously or unconsciously, and how they may be subsequently abandoned. Beliefs about the natural environment and the supernatural environment are part of a larger system of learned ideas and customs that comprise a culture. And entire cultures change over time in the same way beliefs do. Human populations no longer adapt to environmental change by evolving genetically. We now adapt to change by altering our beliefs and behavior. Human evolution from Australopithecines to Cro-Magnon, the first of our ancestors to have the physical features that characterize all human populations today, depended largely on slowly changing gene frequencies. For at least the past 60,000 years, human biological evolution has

Anthony Layng's "Supernatural Power and Cultural Evolution" originally appeared in the *Skeptical Inquirer* 24, no. 6 (November/December 2000).

been relatively inactive. During this time, the evolution of beliefs has come to determine which populations had the greatest capacity to survive.

Cultural and Group Beliefs

At an early stage in cultural evolution, all societies began to believe in supernatural power, the common denominator of all religious beliefs. Tribal societies studied by anthropologists have provided numerous examples of belief systems that may be similar to those of our ancient ancestors. Modern American proclivities to endorse many forms of supernatural power may make us more intellectually similar to tribal societies than to other industrialized ones, but all human cultures today include faith in some spiritual beings and forces. From an evolutionary perspective, this universality suggests that such beliefs must have played some essential role in ensuring the well-being of human populations. And here the reference is not to miracles, magical cures, and spiritual intervention. It is the belief in such things that is likely to be instrumental in this regard, not the things believed in.

Individuals and small groups are capable of cherishing beliefs that are detrimental to their physical welfare, as exemplified by the followers of Heaven's Gate in California or Jim Jones in Guyana. Cults often begin by promoting behavior that is antithetical to the physical well-being of members, but those that evolve into established denominations do so by becoming less exclusive and more reflective of the surrounding culture. Consequently, they abandon their self-destructive behavior. Individuals and fringe groups may hold religious beliefs that compromise their survival, but beliefs that are traditional and generalized throughout a society are likely to enhance the survival of that society. It is as if individuals can afford to be really stupid, but the collective wisdom of a society must be far more practical in its consequence.

Some populations maintain customs that threaten the survival of certain individuals and, at the same time, help to ensure the survival of the society. For example, the belief that firstborn females are cursed and the practice of female infanticide in a tribal society can effectively check population expansion in an environment where unlimited growth would likely lead to mass starvation. Similarly, the traditional prohibition against slaughtering cows in India, where famine was not unusual, meant that many poor people would starve rather than eat beef. But the belief that cows were sacred enabled them and the oxen they produced to survive, thus ensuring that beasts of burden would be available for their essential role in this agrarian society. Had the consumption of beef not been taboo in this largely impoverished and rapidly expanding population, cows would be consumed faster than they could reproduce, oxen would no longer be available for plowing, agricultural output would steadily decline, and the entire society would be hard-pressed to survive.

Proposing that traditional faith in supernatural power has utility for the believers is not at all new. Sigmund Freud, who did not much care for religion, nevertheless gave it credit for its emotional functions such as reducing anxieties. And Emile Durkheim, regarded as the first sociologist, proposed that belief in magic and spiritual beings was highly beneficial for the maintenance of society. But anthropologists now are suggesting that belief in supernatural power is more practical than meeting mere emotional and social needs; it may have played an essential role in ensuring that our ancestors' material needs were met.

Native American Beliefs

To illustrate how belief in supernatural power is capable of providing a reliable supply of scarce goods and resources that sustain a population, I offer the following account of Native American religious beliefs, beginning with the nineteenth-century Plains Indians. According to their tradition, success in hunting, warfare, and other important activities was dependent upon the availability of cosmic supernatural power, energy that was inclined to exhaustion over the course of a year or so. Regenerating this finite power required that the entire tribe assemble and perform an elaborate renewal ritual that took several days.

During most of the year, these hunter and gatherer tribes adapted to their environment by dispersing in numerous bands over a large area. Since the bison herds were similarly distributed, as were other food animals and plants, permanently living as a congregated tribe would cause inevitable starvation. However, the bison gathered annually for a mass migration, and at this time a large coordinated hunt by all the warriors of the tribe could usually furnish enough meat to sustain them through the winter when other food was very difficult to acquire.

The renewal rituals, the most famous of which is the Sundance, occurred at the same time that the bison were assembling, ideal timing for the annual hunt. In addition, this seasonal ritual maintained a sense of tribal identity and loyalty, emotionally tying the various bands together, giving them a collective sense of interdependency and common fate, since the absence of even one band meant that the ritual would not be able to regenerate sufficient power. Since these egalitarian societies were in competition for the limited food supply, and only those that were capable of exercising considerable military strength were in a position to be competitive, the tribes whose bands were strongly unified were the most likely to survive in this politically hostile environment.

When the various Plains Indian tribes were settled on reservations, and bison hunting and intertribal warfare ceased, the annual rituals ceased as well, along with other religious rituals such as vision quests and sweat lodge ceremonies. However, even where many converted to Christianity, a traditional belief in witchcraft survived. Now that tribes were politically dependent and their food was furnished by the government, the annual renewal ritual lost its meaning. But faith in witchcraft remained relevant, because it provided an incentive system to maximize sharing between households—strategically important in a place where food and employment were so scarce.

With a belief in evil clandestine witches (in contrast to those promoted today by feminist spirituality groups), an individual with some money or food was inclined to share it with neighbors out of fear that one of them might be a jealous witch, a man or woman who could cast a spell causing the selfish offender some grievous harm. And since witches were believed to be greedy, as well as jealous, households with surplus goods engaged in overt sharing in order to avoid accusations of witchcraft. Rumors about an individual's alleged sorcery could make that person a social pariah. Many were inclined to share because doing so was in conformance with a strong sharing ethic, but for those who needed additional prodding, the fear of witchcraft and not wanting neighbors to suspect you of witchcraft was likely to ensure that most would share.

Eventually, where poverty was much reduced by economic development, belief in witches declined also. Where education and employment have become viable alternatives to poverty, the witchcraft incentive system that maintains generalized sharing is no longer adaptive. When all are poor, sharing makes possible some minimal security, but a regular income devoted to one's own household can provide even greater security. Individuals might still be accused of being witches, but such accusations in an economically improved environment are likely to be dismissed by residents who are doing relatively well as mere jealousy.

Now that there are tribal colleges and increasing employment opportunities on many reservations, a renewed interest in traditional rituals has emerged. Belief in witches is fading, but Sundances, vision quests, and sweat lodge ceremonies are making a comeback, along with the belief that these rituals can affect supernatural beings and forces. How might this be adaptive among people who are educated?

This interest in traditional religion comes at a time when preserving the tribal language and the old customs is difficult and no longer assured, but one can retain an Indian identity by participating in the ancient rituals. Overt expressions of ethnic identity counterbalance the view that Indians have acculturated to the point where reservations and numerous programs with benefits and payrolls for Indians are no longer warranted. Many in the government favor cutting the federal budget that supports the Bureau of Indian Affairs, special scholarships, subsidized housing, comprehensive health care, and so forth that Native Americans very much rely on. If Indians are no different from other Americans, why maintain this special relationship with the federal government? Rituals like pow-wows and Sundances provide overt evidence that Indians are still Indians, in spite of their casinos and mainstream lifestyle, helping to justify their special political status and the material security that this status provides.

Another traditional religious concept has gained urgency in recent decades, the claim that certain places and properties are sacred. Since most Indians are Christians and have formal education, why all this concerted interest in sacred lands? How might this be adaptive? Consider the fact that the American public has developed considerable sympathy for Indian ideas about land, and that many successful legal land claim cases have proven to be highly lucrative; some tribes have been granted large tracts of valuable real estate, and others have been compensated with millions of dollars by the courts. Tribal lawyers have given ancient ideas about land a contemporary meaning. It seems quite likely that judges in these cases are more inclined to be generous when the "lost" lands in question are "sacred." Here again, it may be the economic advantages that are at the heart of the matter.

When most Native Americans stopped practicing tribal rituals, scholars came to view this as a "loss" of Indian culture; more recently, when they reestablished many of their traditional ceremonies, it was described usually as cultural "revival." Such terms as "loss" and "revival" of culture are descriptive but lack explanatory capacity. To account for the dynamic nature of religious beliefs and practices, we should view them from an evolutionary perspective, recognizing how beliefs that are adaptive tend to be preserved, and how ideas that lose their utility tend to moderate, be reinterpreted, or even disappear.

Modern versus Primitive Societies

Primitive societies with stable social environments are, understandably, resistant to cultural change. Since their beliefs and traditions have adapted to an environment that is relatively constant, their religion will reflect cultural conservatism, suitable for preserving the status quo. For example, they may stress adherence to tradition by being convinced that deified ancestors insist on the old ways. Similarly, their mythology, taken as literal truth, encourages the same conservatism by illustrating how terrible things can happen when the gods are displeased by individuals who break the mores. Belief in witchcraft, as illustrated above, also discourages deviance from traditional norms.

Isolated primitive societies need unchallenged faith in supernatural power. Unfettered rational thinking, objectivity, and skepticism cannot be tolerated, since such challenges to tradition are likely to lead to social change, a very risky development under such circumstances. If they subjected religious belief to scientific verification, if they accepted only what could be observed and experienced, their faith would be so undermined as to be ineffectual in maximizing the availability of essential material resources. Contemporary Americans, on the other hand, can retain or challenge belief in supernatural power without threatening the longevity of their society. The natural resources they depend on are controlled by bureaucratic government, industrial corporations, complex technology, and other secular means.

Industrial countries, in contrast to tradition-oriented societies, rely on flexible forms of social control, such as public opinion and a jury system. This is suitable (adaptive) because the social environment is constantly changing, thus requiring functional equivalents to belief in supernatural power. For example, once it became common practice for conjugal couples to delay or forgo marriage, we changed laws that discriminated against individuals for "living in sin." Similarly, rather than depending exclusively on faith healing and magical cures, most of us now rely on physicians. We have not abandoned religious methods to encourage conformity, and many Americans with severe health problems continue to rely both on traditional prayer and modern surgery. But the survival of our society no longer requires consensus on the idea that supernatural beings and forces sustain us. We, unlike isolated tribal societies, can afford to be skeptical and to encourage rational thinking.

So what might we conclude about belief in supernatural power? What is the most likely explanation for why our early ancestors came to believe in supernatural power? Since this idea turned out to be so valuable for primitive societies, is it reasonable to assume that the belief came from revelation, from the supernatural world? Most people would answer yes because, if true, it seems to confirm that supernatural beings actually exist, and that is what a majority of people believe. But, of course, an explanatory theory cannot be validated by a show of hands.

It is more scientific to consider the possibility that, as the human brain evolved, there was selection for those who were inclined to believe in supernatural power, resulting in a genetic propensity to so believe. Some scholars, inspired by evolutionary psychology, are promoting this theory, but it seems to imply that today's atheists have brains too regressive to recognize what for "normal" people is obvious, that believing in God is just natural. This inborn belief theory has some intellectual appeal, but it lacks hard evidence, as do theological explanations for belief in supernatural power.

Admittedly, at one point in our evolution, human brains became capable of conceptualizing supernatural power, an ability that requires considerable capacity for abstract thought. And, again, given the important adaptive potential of belief in supernatural power, its capacity for aiding the physical survival of a population, early human societies armed with this idea would have been the most likely to last. Over time, with only a Stone Age technology, all surviving societies would exemplify this belief.

This analysis is neither politically correct nor likely to be intellectually palatable to most Americans, for it renders irrelevant the existence of supernatural power and the spiritual beings (God, angels, saints, Satan, etc.) who personify such power. For those, however, who are willing to subject religious convictions to scientific investigation, viewing such concepts from an evolutionary perspective may be highly instructive in suggesting how and why belief in supernatural power emerged and helped to determine the course of cultural evolution.

32
THE BIOLOGICAL ROOTS
OF RELIGION

MORTON HUNT

Why are atheists so different from the overwhelming majority of humankind? Why don't they need to believe in a god of any traditional sort—and most of them not even in a primary force who merely lit the fuse of the Big Bang and then let everything take its own course?

Are they simply more intelligent than almost everybody else? I'm willing to believe they're smarter and more knowledgeable about reality than club-wielding hunter-gatherers, or the members of the Christian Coalition. But can I suppose they're more intelligent than such profoundly religious believers as Plato, Augustine, Aquinas, Descartes, Newton, William James, or even Einstein? Why do a majority of today's American scientists, according to surveys, *not* profess some kind of religious belief?[1]

But the obverse of my puzzlement is far more mystifying: Why have nearly all human beings in every known culture believed in God or gods and accepted the customs, dogmas, and institutional apparatus of an immense array of different religions?

Morton Hunt's "The Biological Roots of Religion" originally appeared in *Free Inquiry* 19, no. 3 (summer 1999).

Belief without Evidence

What makes this so strange is that we human beings have survived, multiplied, and come to dominate the Earth by virtue of our innate tendency to solve problems by taking note of cause-and-effect relationships and making use of them—by observing and using empirical data ranging from the superior flight of an arrow when feathered to the extraordinary expansion of our cognitive powers achieved with computers.

Yet while this indicates that the human mind is basically pragmatic, nearly every human being during recorded history (and to judge from archeological evidence much of prehistory) has held religious beliefs based on no empirical evidence whatever. To be sure, our ancestors of the Homeric and Pentateuchal era often thought they heard the gods talking to them in their minds and sometimes thought they saw them, and even today some mentally ill people, and others who are technically sane but exceedingly pietistic, think they hear God speaking to them or see some fleeting divine apparition. But the great majority of believers neither hear nor see such things. While many sometimes experience a surge of feeling in touch with the divine, the world's believers see not their gods but idols, symbols, and documents representing or telling about their gods.

What other evidence might there be? Many kinds, all highly dubious; real-world events interpreted as God's handiwork can almost always be explained in commonsense or scientific terms. Moreover, the occurrences of miraculous events are almost never weighed against the occurrences of comparable nonevents. We often read in the news of some adorable child dying of inoperable cancer who was marvelously cured when the whole town prayed—but never of the cases in which equally fervent praying did not save the lives of equally adorable children. Nobody remembers them, because human beings have a tendency toward "confirmation bias," as psychologists call it—we remember events that confirm our beliefs but forget those that do not, which is probably why 69 percent of adults in a recent poll said they believe in miracles.[2]

Although realistic knowledge of cause-and-effect relationships has been accumulating over the three centuries of the era of science, it has not eliminated religion. Some believers modify their beliefs to accommodate that evidence, while others reinterpret it most extraordinarily (the fundamentalists say that the geological and fossil traces of Earth's history and of evolution were made by God and planted in the ground during the six days of Creation).

Religion has survived the vast expansion of scientific knowledge by adaptation; except for fundamentalism, it has minimized explaining in supernatural terms whatever can be better explained in natural ones and focused instead on phenomena that cannot be tested or disproved, such as God's mercy, the existence of soul, and the afterlife. Accordingly, more than 90 percent of American adults still believe in God or some form of Higher Being, a large minority have experienced the feeling of being born again,[3] and only 10 percent hold a view of evolution in which God plays no part.[4]

Why, to repeat my central question, do people need religion?

God and Sociobiology

An answer I find persuasive, congruous with historical and social-scientific evidence, and parsimonious is given by sociobiology, the new branch of human behavioral science popularized in 1975 by Edward O. Wilson of Harvard University and now offered in many universities. (In what follows, I draw primarily on three of Wilson's books and on a recent sociobiological study of religion by Professor Walter Burkert of the University of Zurich.[5])

Sociobiology holds that in considerable part human behavior is based on our biology—specifically, by gene-directed tendencies developed in us by evolution. We eat, sleep, build shelters, make love, fight, and rear our young in a wide variety of human fashions because, sociobiologists say, through the process of natural selection interacting with social influences we developed genetic predispositions to behave in ways that ensured our survival as a species. Complex interactions among numerous genes give us the *capacity* and *inclination* to develop into people who are either more or less violent, more or less altruistic, monogamous or polygamous, Muslim or Catholic, or whatever—depending on how our upbringing, experiences, and the myriad influences on us of the culture we are immersed in elicit the potentialities within those congeries of genes.

That's how the individual develops. But how did we come to have a genome that incorporates such developmental possibilities? That's where Wilson's theory comes in. His latest version of his theory centers on what he calls "gene-culture coevolution." He proposes that certain physiologically based preferences channel the development of culture (an example might be the development in every society of some form of family life in response to the infant's and mother's need for continuing sustenance and protection). On the other hand, certain cultural influences reciprocally favor the selection and evolution of particular genetic tendencies (an example might be society's inhibition of uncontrolled aggression and its favoring of people with built-in responsiveness to social control of aggression).

To see how interaction works, consider the case of language. (This is my example, not Wilson's.) No other animal has anything remotely like our language capacity. That's because only the human brain has two specialized zones, Broca's Area and Wernicke's Area, both on the left side, in which the neurons are so connected as to form a mechanism that recognizes the relationships among the words in sentences. No actual language is prewired in those areas; no child, raised apart from the sound of language, has ever spontaneously spoken. But our brains evolved in such a way that every normal toddler can spontaneously figure out what people around him or her are saying, no matter what words and grammar they are using. The evidence of prehistoric skull sizes and shapes, ancient artifacts, and the customs of primitive peoples indicates that the immense advantages of linguistic communication favored individuals with greater neurological capacity for verbal communication, and that culture and genetics coevolved to produce the modern human brain and the resultant thousands of human languages.

This is a paradigm for the development of religion. As Professor Burkert puts it: "We may view religion, parallel to language . . . , as a long-lived hybrid between cultural and the biological traditions."[6] He maintains that we have biological tendencies and capacities that cause us to need, learn, value, and practice religion—not any specific religion, of course, but any one of the thousands of religions that, despite the vast differences among them, all tend to fulfill similar needed functions for individuals and, just as important, for the society they live in.

The primary needs met by religion, sociobiologists say, were the allaying of fear and the explanation of the world's many mystifying phenomena. With the development of the brain's capacity for language, human beings were able to develop concepts and have experiences that had been unavailable to prehumans, among them the consciousness of risk and of death, of time, the past, and the future; of reward and punishment; puzzlement about natural phenomena; the satisfactions of problem solving; and aesthetic pleasure, wonder, and awe.

But verbal and conceptual ability also had rich rewards. Primitive humans developed a sense of awe at the wonders they could now think about: birth, the return of life in spring, the rainbow—and with that sense of awe came a need to explain those wonders. Human beings' new cognitive powers yielded the joys of recognizing health returning after sickness, hardships survived, crops harvested, problems solved, wrongs righted, and the aesthetic pleasure yielded by the many beauties of the world around them.

Early humans, and most humans to this day, make sense of all these mystifying negative and positive experiences by means of religion.

If there is evil in the world, it is, in some religions, the work of an evil deity—Ahriman, Satan, Asmodeus, Loki—but in other religions, it is the product of evil desires in human beings. Against the uncertainties and dangers of the future, people pray, asking the deity to make all turn out well. Against the misery of losing a loved one or the fear of one's own death, people seek reassurance that they will live after death in some other realm. Against injustice, inequality, the desperate unfairness of life, what better consolation than to expect a just and generous reward in heaven by a loving Father? And conversely, when things go well, when the world is beautiful, when people are surrounded by those they love and enjoy the rewards of their work, what is more natural than to give heartfelt thanks to the supposed source of good things?

Religion thus met the newly evolving human need to understand and control life. Religion serves the same purposes as science and the arts—"the extraction of order from the mysteries of the material world," as Wilson puts it[7]—but in the prescientific era there was no other source of order except for philosophy, which was comprehensible only to a favored few and in any case was nowhere nearly as emotionally satisfying as religion.

Still another major function of religion was to act as a binding and cementing social force. I quote Wilson again: "Religion is . . . empowered mightily by its principal ally, tribalism. The shamans and priests implore us in somber cadence, *Trust in the sacred rituals, become part of the immortal force, you are one of us.*"[8] Religious propitiation and sacrifice—near-universals of religious practice—are acts of submission to a dominant being and dominance hierarchy.

Religion thus helped meet the need of human beings to live together. That need is biologically based: We require social life to thrive emotionally—and, in fact, physically. Recent evidence shows that people who live alone have less immune resistance to disease than people who live with spouses or partners. But social living requires some system of hierarchical leadership in order to avoid endless fighting over food, sex, and other benefits. You've seen all this on television documentaries of life among troops of chimpanzees and baboons.

The human creation of various systems of social control is a response to biological urges we inherit from our prehuman ancestors.

But early peoples were aware that certain inexplicable and mighty forces —earthquakes, drought, epidemics—that affected their lives were beyond the control of their leaders. It was only natural that they should suppose that these forces were the work of unseen things analogous to their leaders but far more powerful, and whom they regarded with fear, awe, and respect. From early times to the present, in nearly every religion, God or the gods are the "lords" of creation, rulers whom all humans, including emperors and presidents, must obey and revere. So in addition to whatever form of social governance and leadership human beings developed, they also sought the leadership and help of shamans, medicine men, priests, or other special people who could mediate between them and the spirits or gods, and adopted acts of submission ritual to placate and please those deities. But of course these religious beliefs and practices relieved the leaders of society of the blame when things went wrong; religion thus bolstered social governance.

For all these reasons, says Wilson, "Acceptance of the supernatural conveyed a great advantage throughout prehistory, when the brain was evolving." The human mind evolved to believe in the gods even as religious institutions became built-ins of society.[9]

Inferential Evidence

Although biologists have been able to pinpoint a few genes responsible for certain specific disorders, the genetic basis of any specific form of human behavior is almost certainly due not to a single gene but the intricate interplay of numerous genes. Which ones, however, is still largely undetermined, although it seems certain that in time the details will be spelled out.

The evidence sociobiologists offer is inferential—a set of reasonable and persuasive deductions from what we know about human evolution, human mental abilities, and early religions, including such preliterate evidence as the ceremonial burial objects and wall drawings of Neanderthals and Cro-Magnons. Sociobiologists say that all this evidence strongly supports their theory of religion, for since no other living species exhibits any such behavior, religion must have been a product of evolving human biological traits.

But Burkert says that biological roots of religion are even deeper than, and predate, language, though gaining power and richness when language arrives. One is the device used by many animals of sacrificing a part of themselves in order to escape from danger. Some spiders' legs break off easily and continue to twitch for a while to distract a predator while the spider escapes. Lizards' tails snap off easily, remaining in the grip of the attacker while the lizard makes a getaway and grows a new tail. Some birds, under attack, suddenly shed a mass of feathers, leaving the attacker with a mouthful of fluff while the expected meal disappears.

Human analogs of this behavior exist as religious rituals—sacrifices of desirable possessions to the gods in order to escape ill fortune, such as pouring wine on the ground, slaughtering and burning a valuable animal, giving money to help build a temple. And there are many examples of far more serious sacrifices performed to placate God, such as the self-castrations performed by certain devout early Christians and by the Skoptsi, seventeenth-century Russian religious fanatics. And giving up sexual activity altogether, along with parenthood and family life, as priests and nuns have done for centuries, is surely as extreme a sacrifice of the part for the whole as physical mutilation.

Thus, biology is the basis of the many ritualistic submission acts in human religions. The most general such act, relatively innocuous, is to bend or to bow.[10] Muslims prostrate themselves on the floor; Catholics and some Protestants kneel in prayer; people of nearly all denominations bow their heads submissively in prayer or meditation. Some worshipers beat their chests, weep and cry out, tear their clothes and throw ashes on themselves, crawl for miles on their hands and knees, lash their naked bodies with chains. Even these observances are small potatoes compared to the nauseating acts of devotion of many medieval saints.

A more tasteful genre of biologically based religious behaviors concern cleanliness. Keeping the body clean is a basic necessity for all higher animals, some of whom bathe, others preen, still others groom each other, for the benefit of their bodily functions.[11] We human beings, too, have always taken care of our persons, bathing, cutting our hair, shaving, and so on.

But being human, we conceive of another and far worse kind of dirt that pollutes us: the impurity of wrongdoing. Our ancient ancestors cleansed themselves of wrongdoing through rituals such as burnt offerings, prayer, and self-imposed hardships and humiliations. The Christians improved greatly on all this: They transformed simple guilt for wrongdoing into sin inherited, willynilly, from Adam and Eve. This created a whole new religious industry made up of confession, penance, absolution, communion, and the striving for a cleansed and perfect state, all of which was self-sustaining, since the cleansed person, if normal, was bound to become morally dirty again in a little while.

And so, to sum up the sociobiological theory of the roots of religion: genetically built into early human beings was a set of mental, emotional, and social needs that caused culture to develop in certain ways—including the development of various religions—and caused culture, reciprocally, to favor and select for evolution those human traits that provided sociocultural advantages to the individuals possessing them. "Religion," says Burkert, "follows in the tracks of biology . . . [and] the aboriginal invention of language . . . yield[ing] coherence, stability, and control within this world. This is what the individual is groping for, gladly accepting the existence of nonobvious entities or even principles."[12]

The Unbeliever Puzzle

I return to the first of my puzzlements—Why are unbelievers different from the great majority of their fellow human beings? They are not, however, unique, for throughout civilized history a small minority have not needed supernatural religious explanations of their own thoughts or of the mysteries, tragedies, and glories of everyday life. I refer not just to out-and-out atheists but to that larger minority who have held or hold a deistic concept of God or who regard the inherently consistent laws of nature, governing the behavior of galaxies, genes, and quarks, with the awe and respect that others accord to a more traditional God.

The best example of such a person actually predates modern science. It is Spinoza, for whom God was coterminous with the actual universe, neither outside it nor above it but identical with it and with all natural laws. For him, God was nothing more nor less than the total corpus of those laws.

Perhaps current unbelievers are all contemporary Spinozists, sensitive to and in tune with the god who pervades the universe—who *is* the universe—who is identical with reality. Perhaps unbelievers do not so much reject the religious needs and impulses of the human race as adapt to them in realistic and humanistic terms, replacing the fairy tales of conventional religions with the more intellectually demanding tales, provided by modern science, of natural laws and of the demonstrable, replicable evidence of cause-and-effect relationships.

Perhaps unbelievers meet the basic human need for order and social integration within the subsociety of science itself and its hierarchical structure. Perhaps for unbelievers scientific humanism offers deeply satisfying answers to all those profound and troubling mysteries that religion purports to answer, and unbelievers are comfortable with those answers although they are incomplete and, no matter how our knowledge increases, will remain so, with new discoveries always raising new and more complex questions about reality.

Finally, perhaps unbelievers differ from the great majority of human beings in one other way: possibly unbelievers are psychologically adult, needing no invisible parent figure, able to face the reality of human life and death without fear (or at least live with that fear), and too sensible to believe in anything that has no proof, any explanation of the world that is either *impossibile* or *absurdum*.

But that's only a guess; perhaps I flatter unbelievers unreasonably; perhaps they're not that special and wonderful. But I hope they are.

Notes

1. A 1996 survey quoted in E. O. Wilson, *Consilience* (New York: Alfred A. Knopf, 1998), ca. p. 246.

2. *Time*, April 10, 1995, p. 65.

3. Edward O. Wilson, *On Human Nature* (New York: Bantam, 1979), pp. 176–77.

4. *Freethought Events and Planning Guide*, November 29, 1998.

5. Walter Burkert, *Creation of the Sacred: Tracks of Biology in Early Religions* (Cambridge: Harvard University Press, 1996).

6. Burkert, *Creation of the Sacred*, p. 20.

7. Wilson, *Consilience*, p. 257.

8. Ibid.

9. Ibid., p. 262.

10. Burkert, *Creation of the Sacred*, pp. 84–87.

11. Ibid., p. 123.

12. Ibid., p. 177.

33
WHENCE RELIGIOUS BELIEF?

STEVEN PINKER

"The most common of all follies," wrote H. L. Mencken, "is to believe passionately in the palpably not true. It is the chief occupation of mankind." In culture after culture, people believe that the soul lives on after death, that rituals can change the physical world and divine the truth, and that illness and misfortune are caused and alleviated by spirits, ghosts, saints, fairies, angels, demons, cherubim, djinns, devils, and gods. According to polls, more than a quarter of today's Americans believe in witches, almost half believe in ghosts, half believe in the devil, half believe that the book of Genesis is literally true, 69 percent believe in angels, 87 percent believe that Jesus was raised from the dead, and more than 90 percent believe in a God or universal spirit.

How does religion fit into a mind that one might have thought was designed to reject the palpably not true? The common answer—that people take comfort in the thought of a benevolent shepherd, a universal plan, or an afterlife—is unsatisfying, because it only raises the question of why a mind

Steven Pinker's "Whence Religious Belief?" earlier appeared in the *Skeptical Inquirer* 23, no. 4 (July/August 1999), and is excerpted from his book *How the Mind Works* (New York: W. W. Norton, 1997). Copyright © 1997 by Steven Pinker. Used by permission of W. W. Norton & Company, Inc.

would evolve to find comfort in beliefs it can plainly see are false. A freezing person finds no comfort in believing he is warm; a person face-to-face with a lion is not put at ease by the conviction that it is a rabbit.

What is religion? The psychology of religion has been muddied by scholars' attempts to exalt it while understanding it. Religion cannot be equated with our higher, spiritual, humane, ethical yearnings (though it sometimes overlaps with them). The Bible contains instructions for genocide, rape, and the destruction of families, and even the Ten Commandments, read in context, prohibit murder, lying, and theft only within the tribe, not against outsiders. Religions have given us stonings, witch-burnings, crusades, inquisitions, jihads, fatwas, suicide bombers, abortion-clinic gunmen, and mothers who drown their sons so they can be happily reunited in heaven. As Blaise Pascal wrote, "Men never do evil so completely and cheerfully as when they do it from religious conviction."

Religion is not a single topic. What we call religion in the modern West is an alternative culture of laws and customs that survived alongside those of the nation-state because of accidents of European history. Religions, like other cultures, have produced great art, philosophy, and law, but their customs, like those of other cultures, often serve the interests of the people who promulgate them. Ancestor worship must be an appealing idea to people who are about to become ancestors. As one's days dwindle, life begins to shift from an iterative prisoner's dilemma, in which defection can be punished and cooperation rewarded, to a one-shot prisoner's dilemma, in which enforcement is impossible. If you can convince your children that your soul will live on and watch over their affairs, they are less emboldened to defect while you are alive. Food taboos keep members of the tribe from becoming intimate with outsiders. Rites of passage demarcate the people who are entitled to the privileges of social categories (fetus or family member, child or adult, single or married) so as to preempt endless haggling over gray areas. Painful initiations weed out anyone who wants the benefits of membership without being committed to paying the costs. Witches are often mothers-in-law and other inconvenient people. Shamans and priests are Wizards of Oz who use special effects, from sleight-of-hand and ventriloquism to sumptuous temples and cathedrals, to convince others that they are privy to forces of power and wonder.

Let's focus on the truly distinctive part of the psychology of religion. The anthropologist Ruth Benedict first pointed out the common thread of religious practice in all cultures: religion is a technique for success. Ambrose Bierce defined *to pray* as "to ask that the laws of the universe be annulled on behalf of a single petitioner confessedly unworthy." People everywhere beseech gods and spirits for recovery from illness, for success in love or on the battlefield, and for good weather. Religion is a desperate measure that people resort to when the stakes are high and they have exhausted the usual techniques for the causation of success—medicines, strategies, courtship, and, in the case of the weather, nothing.

What kind of mind would do something as useless as inventing ghosts and bribing them for good weather? How does that fit into the idea that reasoning comes from a system of modules designed to figure out how the world works? The anthropologists Pascal Boyer and Dan Sperber have shown that it fits rather well. First, nonliterate peoples are not psychotic hallucinators who are unable to distinguish fantasy from reality. They know there is a humdrum world of people and objects driven by the usual laws, and find the ghosts and spirits of their belief system to be terrifying and fascinating precisely *because* they violate their own ordinary intuitions about the world.

Second, the spirits, talismans, seers, and other sacred entities are never invented out of whole cloth. People take a construct from one of the cognitive modules—an object, person, animal, natural substance, or artifact—and cross out a property or write in a new one, letting the construct keep the rest of its standard-issue traits. A tool or weapon or substance will be granted some extra causal power but otherwise is expected to behave as it did before. It lives at one place at one time, is unable to pass through solid objects, and so on. A spirit is stipulated to be exempt from one or more of the laws of biology (growing, aging, dying), physics (solidity, visibility, causation by contact), or psychology (thoughts and desires are known only through behavior). But otherwise the spirit is recognizable as a kind of person or animal. Spirits see and hear, have a memory, have beliefs and desires, act on conditions that they believe will bring about a desired effect, make decisions, and issue threats and bargains. When the elders spread religious beliefs, they never bother to spell out these defaults. No one ever says, "If the spirits promise us good weather in exchange for a sacrifice, and they know we want good weather, they predict that we will make the sacrifice." They don't have to, because they know that the minds of the pupils will automatically supply these beliefs from their tacit knowledge of psychology. Believers also avoid working out the strange logical consequences of these piecemeal revisions of ordinary things.

They don't pause to wonder why a God who knows our intentions has to listen to our prayers, or how a God can both see into the future and care about how we choose to act. Compared to the mind-bending ideas of modern science, religious beliefs are notable for their lack of imagination (God is a jealous man; heaven and hell are places; souls are people who have sprouted wings). That is because religious concepts are human concepts with a few emendations that make them wondrous and a longer list of standard traits that make them sensible to our ordinary ways of knowing.

But where do people get the emendations? Even when all else has failed, why would they waste time spinning ideas and practices that are useless, even harmful? Why don't they accept that human knowledge and power have limits and conserve their thoughts for domains in which they can do some good? I have alluded to one possibility: the demand for miracles creates a market that would-be priests compete in, and they can succeed by exploiting people's dependence on experts. I let the dentist drill my teeth and the surgeon cut into my body even though I cannot possibly verify for myself the assumptions they use to justify those mutilations. That same trust would have made me submit to medical quackery a century ago and to a witch doctor's charms millennia ago. Of course, witch doctors must have *some* track record or they would lose all credibility, and they do blend their hocus-pocus with genuine practical knowledge such as herbal remedies and predictions of events (for instance, the weather) that are more accurate than chance.

And beliefs about a world of spirits do not come from nowhere. They are hypotheses intended to explain certain data that stymie our everyday theories. Edward Tylor, an early anthropologist, noted that animistic beliefs are grounded in universal experiences. When people dream, their body stays in bed but some other part of them is up and about in the world. The soul and the body also part company in the trance brought on by an illness or a hallucinogen. Even when we are awake, we see shadows and reflections in still water that seem to carry the essence of a person without having mass, volume, or continuity in time and space. And in death the body has lost some invisible force that animates it in life. One theory that brings these facts together is that the soul wanders off when we sleep, lurks in the shadows, looks back at us from the surface of a pond, and leaves the body when we die.

34
SEARCHING FOR GOD IN THE MACHINE

DAVID C. NOELLE

"I have certain, positive knowledge from my own direct experience. I can't put it any plainer than that. I have seen God face to face." With these words, the fictional theologian Palmer Joss defends his religious convictions in Carl Sagan's 1985 novel, *Contact*. Joss argues for the existence of his Christian god on the basis of personal revelation. And Joss is not alone. Many religionists rest their faith on the apparently solid foundation of personal religious experiences. Some receive visions. Others hear a comforting voice. Almost all experience a "sense of presence" or a feeling of "unity with the universe." Such episodes typically bring catharsis, joy, and calm. Importantly, these experiences are not reported solely by people suffering from brain damage or mental illness.

We may, quite rightly, reject such subjective experiences as lacking the necessary qualities of scientific evidence, such as reproducibility and openness to consensual validation or critique. The religionist may retort, however, that his belief may not be scientifically justifiable, but he knows it to be true, nonetheless, because of his private religious revelation.

To completely counter this argument for the existence of god(s), some

David C. Noelle's "Searching for God in the Machine" originally appeared in *Free Inquiry* 18, no. 3 (summer 1998).

alternative explanation must be given for the religious experience. Researchers in the fields of psychology and neuroscience have begun to uncover the biological mechanisms that might give rise to feelings of revelation in healthy adults. I will briefly review and critically assess some of these scientific findings, focusing specifically on three questions:

1. Circuitry—What brain circuits are involved in religious experiences?

2. Modularity—Does the brain contain a special module dedicated to religious experience?

3. Innateness—Is there a "religion instinct" that is genetically "hard-wired" into our brains?

The 'God Module' Discovery

In October 1997 a presentation boldly titled "The Neural Basis of Religious Experience" was given at the annual conference of the Society for Neuroscience by neuropsychologist Dr. V. S. Ramachandran and his colleagues. The *Los Angeles Times* reported that, "researchers at the UC-San Diego brain and perception laboratory determined that the parts of the brain's temporal lobe—which the scientists quickly dubbed the 'God module'—may affect how intensely a person responds to religious beliefs." The story suggested that there were now at least partial scientific answers to all three of our questions concerning religious experience. The circuits underlying religious experience are in the temporal lobe of the brain; they form a distinct religion module that is substantially innate. Experimental results were "leading the scientists to suggest a portion of the brain is naturally attuned to ideas about a supreme being."

One common way to hunt for a module in the brain is to examine patients with various kinds of brain damage, hoping to find a localized form of damage that correlates with changes in the behaviors of interest. In this way, one may discover relationships between certain circuits in the brain and certain behavioral functions.

This was the strategy taken by the scientists in San Diego. They decided to focus on temporal lobe epilepsy (TLE) patients, who exhibit interictal behavior syndrome (IBS). These patients are prone to excessive activity in their temporal lobes, causing seizures during which they report powerful religious experiences. Importantly, clinicians have previously reported that such TLE patients are also often fanatically religious, even during the long periods between seizures.

The question asked by Ramachandran and his colleagues was, Why do such seizures often lead to enhanced religiosity? They entertained three possibilities:

1. Strange sensory experiences that arise during seizure are rationally interpreted as signs of paranormal powers.

2. The strong and widespread electrical activity that defines seizures strengthens connections between temporal lobe sensory areas and the amygdala (a brain area associated with emotion). This causes patients to see "deep cosmic significance" in everything.

3. There is a system in the temporal lobe devoted to mediating emotional responses of a religious nature. Seizures can selectively strengthen the connections in this system.

The researchers dismissed the first option on the grounds that other kinds of neurological and psychiatric disorders result in strange hallucinations without causing the development of specifically religious propensities. To distinguish between the remaining two options, the scientists tested TLE-IBS patients to see if they had stronger emotional responses to everything in the world or only to religious stimuli. The degree of emotional response was measured through a physiological correlate, skin conductance response (SCR). By measuring small rapid changes in perspiration, the researchers hoped to show that TLE-IBS patients were particularly aroused by religion.

Indeed, that was exactly what was found. The TLE-IBS patients showed preferential emotional arousal when presented with religious words as opposed to words with sexual or violent connotations. Unlike the patients, age-matched healthy control subjects responded most strongly to the sexual words.

While these results seem to indicate that there is something distinctly religious about some of the circuitry in the temporal lobes, there are some reasons to be cautious about this conclusion. First, this study involved only three patients, and the preference for religious words was not equally robust in all three. Also, TLE-IBS patients sometimes exhibit changes in sexuality, becoming obsessed with the topic or bored by it. This symptom could have impacted the patients' responses to the sexual stimuli. In short, these results should be seen as preliminary.

In the end, this experiment suggests only that TLE-IBS patients do, indeed, display religion-specific symptoms. This, in turn, suggests that the brain's temporal lobe is involved in religious experience. The degree to which religion is a distinct or genetically determined part of our neural architecture has yet to be determined.

Tuning into the Divine

In the film *Raiders of the Lost Ark*, a large metal religious artifact called the "Ark of the Covenant" is described as "a radio for speaking to God." While, to date, no one has produced such a radio that can tune into the divine, one neuroscientist has fabricated something similar. Dr. Michael Persinger of Laurentian University has devised a machine that generates a particular kind of magnetic field around the head, producing "micro-seizures" in the temporal lobes of the brain. Healthy people who have experienced this induced brain activity have reported such things as a feeling of floating, deformations of the body, strong emotions, a "sensed presence," and specifically religious dreamlike hallucinations.

Persinger's experimental work arose after years of research into the neurological basis of religious experiences. Over this time, he has constructed and refined a rather detailed account of the neural processes that may underlay feelings of supernatural contact.[1] In brief, religious experiences are seen as the result of "temporal lobe transients" (TLT)—short-lived rate increases and instability in the firing patterns of neurons in the temporal lobe. These transients are seen as miniature versions of the seizures experienced by temporal lobe epileptics, and they are thought to occasionally arise in healthy people.

Persinger has speculated as to why such TLT events would produce the particular configuration of experiences reported as religious revelations. He sees a critical part of our "sense of self" as being maintained by systems in the left hemisphere temporal cortex. Most of the time, there is "matched" activity in the analogous places in the right hemisphere. However, when activity on the right gets out of sync with activity on the left, as during a TLT event, the left hemisphere interprets the mismatched activity as "another self" or a "sensed presence"—the mind of God. In conjunction with this experience comes excessive stimulation of subcortical areas in the temporal lobe, particularly the amygdala (associated with emotion) and the hippocampus (associated with autobiographical memory). Excitation of these areas results in the attribution of personal meaning to the experience. These powerful TLT events may naturally result from a number of factors, including increased sensitivity or lability of right temporal areas, loss of oxygen to the brain, and changes in blood sugar. These biological conditions may be caused by crisis situations, prolonged anxiety, near-death contingencies, high altitudes, starvation and fasting, diurnal shifts, and other physiological stressors.

A variety of correlational studies on healthy adults have been conducted by scientists in Persinger's lab. Assuming that people who are prone to TLT events will show subtle signs of a tendency towards hemispheric mismatch even when not experiencing a "micro-seizure," Persinger and his colleagues have examined the "brain waves" of a large number of healthy subjects and have compared these results with reports of religious experiences. They have found that a particular low-frequency component of one's electroencephalogram (EEG) trace, known as the theta rhythm, can partially predict the likelihood of having religious experiences.[2] Across healthy subjects, hemispheric mismatch in the theta component correlates with reports of previous "sensed presence" experiences. Furthermore, signs of specifically subcortical (limbic) mismatch in the temporal lobes are correlated with belief in paranormal phenomena, whereas indications of mismatch in the cortex are correlated with previous "sensed presence" experiences.[3] In short, there is good correlational evidence that one's tendency to have religious experiences involves interhemispheric circuits in the temporal lobe.

While these results are interesting, Persinger's work involving the actual generation of religious experiences is much more striking. In a typical experiment, the subject is isolated from sounds and the eyes are covered. A helmet equipped with solenoids is strapped to the head. While reclining in this state of partial sensory deprivation, currents are induced in the subject's brain through the generation of patterned extremely low frequency milligauss magnetic fields in the solenoids. The subject is asked to describe any experiences aloud, and this monologue is recorded.

By manipulating the magnetic field, the experimenter has some control over the location and pattern of induced current in the brain. When subcortical (limbic) areas in the temporal lobes are targeted, subjects often report distortions in their body images, senses of forced motion, and strong emotional reactions. For example, Dr. Susan Blackmore entered Persinger's experimental chamber and reported a sense of swaying motion, a feeling of being yanked into an upright position, a sense that her leg had been stretched halfway to the ceiling, a period of intense anger, and a flash of terror.[4] When temporal cortical areas are targeted for stimulations, subjects often report dreamlike visions (often with mystical or religious content), a "sense of presence," and strong emotions. Journalist Ian

Cotton, for example, reported highly detailed visions of his childhood home, a dreamlike visit to the monks of a Tibetan temple, and an emotional "realization" that he too was, and always had been, a Tibetan monk.[5] Visions are particularly sensitive to suggestion, with the content being influenced by, say, the presence of a crucifix or the playing of distinctly Eastern music.

With these experimental results in mind, our three questions might be asked of both Persinger's theory and of his data. Persinger holds that subcortical temporal lobe systems contribute to paranormal experiences and paranormal belief. Cortical areas in the temporal lobes participate in the "sense of self" and, during periods of hemispheric mismatch, in the "sensed presence." His correlational and experimental data both support the notion that temporal circuits are central to religious experience. With regard to the question of modularity, Persinger's theory specifically denies the existence of a distinct "God module." In his view, the brain areas responsible for religious experience are exactly those areas that also mediate "sense of self," general emotional responses, and autobiographical memory. While his experimental work does not bear on this question, his correlational data support this distributed view. The likelihood of having religious experiences is systematically related to these other properties of cognition.

Persinger's position on the question of innateness is more ambiguous. In his writings, he frequently points out that religious experiences can have positive effects. He sees TLT events as a remedy for extreme anxiety.[6]

The God Experience has had survival value. It has allowed the human species to live through famine, pestilence, and untold horrors. When temporal lobe transients occurred, men and women who might have sunk into a schizophrenic stupor continued to build, plan, and hope.

While such "survival value" facilitates the incorporation of a feature into the genome, the utility of a behavior is not enough to ensure such fixation in DNA. For example, the making of bread is a skill with great survival value, but it is unlikely that this skill is genetically encoded. Still, Persinger seems to lean towards a largely nativist account. Unfortunately, the data that has emerged from Persinger's lab does not really address the question of innateness.

Note that, even if a tendency towards experiencing TLT events was found to be influenced by one's genes, this would not necessarily mean that religious experiences have been favored by natural selection. For example, it might be the case that temporal lobe lability contributes to imagination and creativity, and this lability also accidentally results in religious experiences. In short, the question of a "religion instinct" is far from settled by Persinger's work.

Persinger's investigations have yet to fully confirm his views on the neurological bases of religious experience, but he has made tremendous progress. Unlike Ramachandran's work with TLE-IBS patients, Persinger has focused on healthy adults. He has shown that particular activity patterns in the temporal lobes of healthy brains can give rise to experiences that are very similar to the spontaneous religious experiences reported by many.

Answering Revelations

Modern science is beginning to understand the neurological mechanisms that give rise to the religious experiences of the believer. Given these results, the skeptic may present the believer with a simple question: How do you know that your religious experience is not a simple trick of your brain—the unfolding of a perfectly natural temporal lobe transient? How can you trust such an experience when, through science, we can convincingly mimic the face of God?

Notes

1. Michael A. Persinger, *Neuropsychological Bases of God Beliefs* (New York: Praeger, 1987).

2. C. Munro and Michael A. Persinger, "Relative Right Temporal-lobe Theta Activity Correlates with Vingiano's Hemispheric Quotient and the 'Sensed Presence,'" *Perceptual and Motor Skills* 75 (1992): 899–903.

3. Michael A. Persinger, "Paranormal and Religious Beliefs May Be Mediated Differently by Subcortical and Cortical Phenomenological Processes of the Temporal (Limbic) Lobes," *Perceptual and Motor Skills* 76 (1993): 247–51.

4. Susan Blackmore, "Alien Abduction: The Inside Story," *New Scientist* 19 (November 1994): 29–31.

5. Ian Cotton, "Dr. Persinger's God Machine," *Free Inquiry* 17, no. 1 (winter 1996/97): 47–51.

6. Persinger, *Neuropsychological Bases of God Beliefs*, p. 138.

VII

Accommodating
Science and Religion

35
SCIENCE AND THE UNKNOWABLE

MARTIN GARDNER

Existence, the preposterous miracle of existence! To whom has the world of opening day never come as an unbelievable sight? And to whom have the stars overhead and the hand and voice nearby never appeared as unutterably wonderful, totally beyond understanding? I know of no great thinker of any land or era who does not regard existence as the mystery of all mysteries.

—John Archibald Wheeler

One of the fundamental conflicts in philosophy, perhaps the most fundamental, is between those who believe that the universe open to our perception and exploration is all there is, and those who regard the universe we know as an extremely small part of an unthinkably vaster reality. These two views were taken by those two giants of ancient Greek philosophy, Plato and Aristotle. Plato, in his famous cave allegory, likened the world we experience to the shadows on the wall of a cave. To turn this into a mathematical metaphor, our universe is like a projection onto three-dimensional space of a much larger realm in a higher space-time.

Martin Gardner's "Science and the Unknowable" originally appeared in the *Skeptical Inquirer* 22, no. 6 (November/December 1998).

For Aristotle the universe we see, although parts of it are beyond human comprehension, is everything. It is a steady-state cosmos, self-caused, having no beginning or end. There is no Platonic realm of transcendent realities and deities. Plato succumbed to what Paul Kurtz likes to call the "transcendental temptation." Aristotle managed to avoid it.

In recent years cosmologists have blurred the distinction between the universe we know and transcendent regions by positing a "multiverse" in which an infinity of universes are continually exploding into existence, each with a unique set of laws and constants. This is one way to defend the anthropic principle against the argument that the universe's fine tuning is evidence of a Designer. It is known that if any of some dozen constants is altered by a minuscule fraction it would not be possible for suns and planets to form, let alone life to evolve. The counterargument: If there is an infinity of universes, each with an unplanned, random set of constants, then obviously we must exist only in a universe with constants that permit life to evolve.

The multiverse concept, however, is far from a step toward Platonic transcendence. The other universes do not differ from ours in any truly fundamental way. They all spring into being in response to random fluctuations in the same laws of quantum mechanics, varying only in the accidental way their Big Bang creates laws. There is still no need to leap from a godless nature to transcendental regions that somehow lie beyond the multiverse.

A few cosmologists and far-out philosophers have gone much further. They conjecture that all possible universes exist—that is, every universe based on a noncontradictory set of laws. In the many-worlds interpretation of quantum mechanics, the universe is constantly splitting into parallel worlds, but these countless worlds all obey the same laws. The multi-multiverse of all-possible-worlds is a much larger ensemble, obviously infinite because the number of logically consistent possibilities is infinite. Most physicists do not buy this view because it is the utmost imaginable violation of Occam's razor. Leibniz's notion of a Creator who surveyed all logically possible worlds, then selected what She considered the most desirable, is surely a simpler conjecture by many orders of magnitude.

A question now arises. As science steadily advances in its knowledge of nature, never reaching absolute certainty but always getting closer and closer to understanding nature, will it eventually discover everything?

We have to be careful to define what is meant by "everything." There is a trivial sense in which humanity cannot possibly know all there is to know. We will never know how many hairs were on Plato's head when he died, or whether Jesus sneezed while delivering the Sermon on the Mount. We will never know all the decimal digits of pi, or all possible theorems of geometry. We will never know all possible theorems just about triangles. We will never know all possible melodies, or poems, or novels, or paintings, or jokes, or magic tricks because the possible combinations are limitless. Moreover, as Kurt Gödel taught us, every mathematical system complex enough to include arithmetic contains theorems that cannot be proved true or false within the system. Whether Gödelian undecidability may apply to mathematical physics is not yet known.

When physicists talk about TOEs (Theories of Everything) they mean something far less trivial. They mean that all the fundamental laws of physics eventually will become known, perhaps unified by a single equation or a small set of equations. If this happens, and physicists find what John Wheeler calls the Holy Grail, it will of course leave unknown billions and billions of questions about the complexities that emerge from the fundamental laws.

At the moment, cosmologists do not know the nature of "dark matter" that holds together galaxies, or how fast the universe is expanding, and hundreds of other unanswered questions. Biologists do not know how life arose on Earth or whether there is life on planets in other solar systems. Evolution is a fact, but deep mysteries remain about how it operates. No one has any idea how complex organic molecules are able to fold so rapidly into the shapes that allow them to perform their functions in living organisms. No one knows how consciousness emerges from the brain's complicated molecular structure. We do not even know how the brain remembers.

Such a list of unknowns could fill a book, but all of them are potentially knowable if humanity survives long enough. Too often in the past scientists have decided that something is permanently unknowable only to be contradicted a few generations later. On the other hand, many scientists have predicted that physics was near the end of its road only to have enormous new revolutions of knowledge take place a few decades later.

In recent years, just when it was thought that all the basic particles had been found or conjectured, along came superstrings, the most likely candidate at the moment for a TOE. If superstring theory is correct, it means that all fundamental particles are made of incredibly tiny loops of enormous tensile strength. The way they vibrate generates the entire zoo of particles.

What are superstrings made of? As far as anyone knows they are not made of anything. They are pure mathematical constructs. If superstrings are the end of the line, then everything that exists in our universe, including you and me, is a mathematical construction. As a friend once said, the universe seems to be made of nothing, yet somehow it manages to exist.

On the other hand, superstrings may, at some future time, turn out to be composed of still smaller entities. Many famous scientists, notably Arthur Stanley Eddington, David Bohm, Eugene Wigner, Freeman Dyson, and Stanislaw Ulam, believed that the universe has bottomless levels. As soon as one level is penetrated, a trapdoor opens to a hitherto unsuspected subbasement. These subbasements are infinite. As the old joke goes, it's turtles all the way down. Here is how Isaac Asimov expressed this opinion in his autobiography, *I, Asimov*: "I believe that scientific knowledge has fractal properties; that no matter how much we learn, whatever is left, however small it may seem, is just as infinitely complex as the whole was to start with. That, I think, is the secret of the Universe."

A similar infinity may go the other way. Our universe may be part of a multiverse, in turn part of a multi-multiverse, and so on without end. As one of H. G. Wells's fantasies has it, our cosmos may be a molecule in a ring on a gigantic hand.

Even if the universe is finite in both directions, and there are no other worlds, are there fundamental questions that can never be answered? The slightest reflection demands a yes.

Suppose that at some future date a TOE will provide all the basic laws and constants. Explanation consists of finding a general law that explains a fact or a less general law. Why does Earth go around the Sun? Because it obeys the laws of gravity. Why are there laws of gravity? Because, Einstein revealed, large masses distort space-time, causing objects to move along geodesic paths. Why do objects take geodesic paths? Because they are the shortest paths through space-time. Why do objects take the shortest paths? Now we hit a stone wall. Time, space, and change are given aspects of reality. You can't define any of these concepts without introducing the concept into the definition. They are not mere aspects of human consciousness, as Kant imagined. They are "out there," independent of you and me. They may be unknowable in the sense that there is no way to explain them by embedding them in more general laws.

Imagine that physicists finally discover all the basic waves and their particles, and all the basic laws, and unite everything in one equation. We can then ask "Why that equation?" It is fashionable now to conjecture that the Big Bang was caused by a random quantum fluctuation in a vacuum devoid of space and time. But of course such a vacuum is a far cry from nothing. There had to be quantum laws to fluctuate. And why are there quantum laws?

Even if quantum mechanics becomes "explained" as part of a deeper theory—call it X—as Einstein believed it eventually would be, then we can ask "Why X?" There is no escape from the superultimate questions: "Why is there something rather than nothing, and why is the something structured the way it is?" As Stephen Hawking recently put it, "Why does the universe go to all the bother of existing?" The question obviously can never be answered, yet it is not emotionally meaningless. Meditating on it can induce what William James called an "ontological wonder-sickness." Jean Paul Sartre called it "nausea." Fortunately such reactions are short-lived or one could go mad by inhaling what James called "the blighting breath of the ultimate why."

Consider the extremely short time humanity has been evolving on our little planet. It seems unlikely that evolution has stopped with us. Can anyone believe that a million years from now, if humanity lasts, that our brains will not have evolved far beyond their present capacities? Our nearest relatives, the chimpanzees, are incapable of understanding why three times three is nine, or anything else taught in grade school. It is difficult to imagine that a million years from now our brains will not be grasping truths about the universe that are as far beyond what we now can know as our understanding is beyond the mind of a monkey. To suppose that our brains, at this stage of an endless process of evolution, are capable of knowing everything that can be known strikes me as the ultimate in hubris.

If one is a theist, obviously there is a vast unknowable reality, transcending our universe, a "wholly other" realm impossible to contemplate without an emotion of what Rudolph Otto called the *mysterium tremendum*. But even if one is an atheist or agnostic, the Unknowable will not go away. No philosopher has written more persuasively about this than agnostic Herbert Spencer in the opening chapters of his *First Principles* (1894).

On the beginning hundred pages of this book, in a part titled "The Unknowable," Spencer argues that a recognition of the Unknowable is the only way to reconcile science with religion. The emotion behind all religions, aside from their obvious superstitions and gross beliefs, is one of awe toward the impenetrable mysteries of the universe. Here is how Spencer reasoned:

One other consideration should not be overlooked—a consideration which students of Science more especially need to have pointed out. Occupied as such are with established truths, and accustomed to regard things not already known as things to be hereafter discovered, they are liable to forget that information, however extensive it may become, can never satisfy inquiry. Positive knowledge does not, and never can, fill the whole region of possible thought. At the uttermost reach of discovery there arises, and must ever arise, the question—What lies beyond? As it is impossible to think of a limit to space so as to exclude the idea of space lying outside that limit; so we cannot conceive of any explanation profound enough to exclude the question—What is the explanation of that explanation? Regarding Science as a gradually increasing sphere, we may say that every addition to its surface does but bring it into wider contact with surrounding nescience. There must ever remain therefore two antithetical modes of mental action. Throughout all future time, as now, the human mind may occupy itself, not only with ascertained phenomena and their relations, but also with that unascertained something which phenomena and their relations imply. Hence if knowledge cannot monopolize consciousness—if it must always continue possible for the mind to dwell upon that which transcends knowledge, then there can never cease to be a place for something of the nature of Religion; since Religion under all its forms is distinguished from everything else in this, that its subject matter passes the sphere of the intellect.

By "Religion" Spencer did not mean religion in the usual sense of worshiping God or gods, but only a sense of awe and wonder toward ultimate mysteries. For him Science and Religion were two essential aspects of thought; Science expressing the knowable, Religion the unknowable. The two merge without contradiction. "If Religion and Science are to be reconciled," he writes, "the basis of reconciliation must be this deepest, widest, most certain of all facts—that the Power which the Universe manifests to us is inscrutable."

No matter how many levels of generalization are made in explaining facts and laws, the levels must necessarily reach a limit beyond which science is powerless to penetrate.

In all directions his investigations eventually bring him face to face with an insoluble enigma; and he ever more clearly perceives it to be an insoluble enigma. He learns at once the greatness and the littleness of the human intellect—its power in dealing with all that comes within the range of experience, its impotence in dealing with all that transcends experience. He, more than any other, truly knows that in its ultimate nature nothing can be known.

The rest of Spencer's First Principles, titled "The Knowable," is an effort to summarize the science of his day, especially what was then known about evolution.

But an account of the Transformation of Things, given in the pages which follow, is simply an orderly presentation of facts; and the interpretation of the facts is nothing more than a statement of the ultimate uniformities they present—the laws to which they conform. Is the reader an atheist? The exposition of these facts and these laws will neither yield support to his belief nor destroy it. Is he a pantheist? The phenomena and the inferences as now to be set forth will not force on him any incongruous implication. Does he think that God is immanent throughout all things, from concentrating nebulae to the thoughts of poets? Then the theory to be put before him contains no disproof of that view. Does he believe in a Deity who has "given unchanging laws to the Universe?" Then he will find nothing at variance with his belief in an exposition of those laws and an account of the results.

Boundaries and Barriers: On the Limits of Scientific Knowledge (Addison-Wesley, 1996), edited by John Casti and Anders Karlqvist, is one of a spate of recent books on the topic. For almost all its authors, the term "limits" is confined to unsolved but potentially solvable questions. Most of the authors agree with what the editors say in their introduction: "Unlike mathematics, there is no knock-down airtight argument to believe that there are questions about the rest of the world that we cannot answer in principle."

Only British astronomer John Barrow has the humility to disagree. He concludes his contribution as follows:

> In this brief survey we have explored some of the ways in which the quest for a Theory of Everything in the third millennium might find itself confronting impassable barriers. We have seen there are limitations imposed by human intellectual capabilities, as well as by the scope of technology. There is no reason why the most fundamental aspects of the laws of nature should be within the grasp of human minds, which evolved for quite different purposes, nor why those laws should have testable consequences at the moderate energies and temperatures that necessarily characterize life-supporting planetary environments. There

are further barriers to the questions we may ask of the universe, and the answers that it can provide us with. These are barriers imposed by the nature of knowledge itself, not by human fallibility or technical limitations. As we probe deeper into the intertwined logical structures that underwrite the nature of reality, we can expect to find more of these deep results which limit what can be known. Ultimately, we may even find that their totality characterizes the universe more precisely than the catalogue of those things that we can know.

Barrow later expanded these sentiments in his 1998 book, *Impossibility: The Limits of Science and the Science of Limits* (Oxford University Press). Here are some passages from his final courageous chapter:

> The idea that some things may be unachievable or unimaginable tends to produce an explosion of knee-jerk reactions amongst scientific (and not so scientific) commentators. Some see it as an affront to the spirit of human inquiry: raising the white flag to the forces of ignorance. Others fear that talk of the impossible plays into the hands of the anti-scientists, airing doubts that should be left unsaid lest they undermine the public perception of science as a never-ending success story. . . .
>
> We live in strange times. We also live in strange places. As we probe deeper into the intertwined logical structures that underwrite the nature of reality, I believe that we can expect to find more of these deep results which limit what can be known. Our knowledge about the Universe has an edge. Ultimately, we may even find that the fractal edge of our knowledge of the Universe defines its character more precisely than its contents; that what cannot be known is more revealing than what can.
>
> George Gamow once described science as an expanding circle, not on a plane but on a sphere. It reaches a maximum size, after which it starts to contract until finally the sphere is covered and no more fundamental knowledge about the universe remains. In recent years numerous physicists, Hawking for instance, have expressed similar hopes. Richard Feynman suggested that although the circle may start to contract, it will become ever more difficult to obtain new knowledge and close the circle completely.
>
> That science will soon discover everything is far from a recent hope. William James, lecturing at Harvard more than a century ago, attacked the hope with these words:

In this very University . . . I have heard more than one teacher say that all the fundamental conceptions of truth have already been found by science, and that the future has only the details of the picture to fill in. But the slightest reflection . . . will suffice to show how barbaric such notions are. They show such a lack of scientific imagination, that it is hard to see how one who is actively advancing any part of science can make a mistake so crude. . . .

Our science is a drop, our ignorance a sea. Whatever else be certain, this at least is certain—that the world of our present natural knowledge is enveloped in a larger world of some sort of whose residual properties we at present can frame no positive idea.

Infinite in All Directions, the title of Freeman Dyson's 1988 book, says it all. Near the close of his third chapter he has this to say about a different hope:

It is my hope that we may be able to prove the world of physics as inexhaustible as the world of mathematics. Some of our colleagues in particle physics think that they are coming close to a complete understanding of the basic laws of nature. They have indeed made wonderful progress in the last ten years. But I hope that the notion of a final statement of the laws of physics will prove as illusory as the notion of a formal decision process for all of mathematics. If it should turn out that the whole of physical reality can be described by a finite set of equations, I would be disappointed. I would feel that the Creator had been uncharacteristically lacking in imagination. I would have to say, as Einstein once said in a similar context, "Da könnt' mir halt der liebe Gott leid tun" ("Then I would have been sorry for the dear Lord").

36
A WAY OF LIFE FOR AGNOSTICS

JAMES LOVELOCK

The naming of things is important. Our deepest thoughts are unconscious, and we need metaphors and similes to translate them into something that we, as well as the rest of humankind, can understand. This is especially true of the broad subject, Gaia theory, which is the pseudonym for Earth System Science. Many scientists seem to dislike Gaia as a name; prominent among them is the eminent evolutionary biologist John Maynard Smith. He made clear when he said of Gaia, "What an awful name to call a theory," that it was the name, the metaphor, more than the science that caused his disapproval. He was, like most scientists, well aware of the power of metaphor. William Hamilton's metaphors of selfish and spiteful genes have served wonderfully, in Richard Dawkins's hands, to make evolutionary science comprehensible, but let us never forget that the powerful metaphor of Gaia was the gift of a great novelist. I would remind those who criticize the name Gaia that they are doing battle with William Golding, who first coined it. We should not lightly turn aside from the name Gaia because of pedantic objection. Biologists now accept Gaia as a theory that they can try to falsify, so why do they continue to object to the name

James Lovelock's "A Way of Life for Agnostics" originally appeared in the *Skeptical Inquirer* 25, no. 5 (September/October 2001).

itself? Surely, it cannot be metaphor envy. I think that it is something deeper, a rejection by reductionist scientists of anything that smells of holism, anything that implies that the whole may be more than the sum of its parts. I see the battle between Gaia and the selfish gene as part of an outdated and pointless war between holists and reductionists. In a sensible world, we need them both.

The philosopher Mary Midgley reminded us that Gaia has influence well beyond science. She said, "The reason why the notion of this enclosing whole concerns us is that it corrects a large and disastrous blind spot in our contemporary worldview. It reminds us that we are not separate, independent autonomous entities. Since the Enlightenment, the deepest moral efforts of our culture have gone to establishing our freedom as individuals. The campaign has produced great results but like all moral campaigns it is one-sided and has serious costs when the wider context is forgotten."

One of these costs is our alienation from the physical world. She went on to say, "We have carefully excluded everything nonhuman from our value system and reduced that system to terms of individual self-interest. We are mystified—as surely no other set of people would be—about how to recognize the claims of the larger whole that surrounds us—the material world of which we are a part. Our moral and physical vocabulary, carefully tailored to the social contract, leaves no language in which to recognize the environmental crisis."

Strangely, a statesman led me to think similar thoughts. That noble and brave man, Vaclav Havel, stirred me to see that science could evolve from its self-imposed reductionist imprisonment. His courage against adversity gave his words authority. When Havel was awarded the Freedom Medal of the United States he took as the title of his acceptance speech, "We are not here for ourselves alone." He reminded us that science had replaced religion as the source of knowledge but that modern science offers no moral guidance. He went on to say that recent holistic science did offer something to fill this moral void. He cited the anthropic principle as explaining why we are here, and Gaia as something to which we could be accountable. If we could revere our planet with the same respect and love that we gave in the past to God, it would benefit us as well as Earth. Perhaps those who have faith might see this is God's will also.

I do not think that President Havel was proposing an alternative Earth-based religion. I take his suggestion as offering something quite different. I think he offered a way of life for agnostics. Gaia is a theory of science and is therefore always provisional and evolving, it is never dogmatic or certain and could even be wrong. Provisional it may be but being of the palpable Earth, it is something tangible to love and fear and think we understand. We can put our trust—even faith—in Gaia but this is different from the cold certainty of purposeless atheism or an unwavering belief in God's purpose.

Science is not excluded from Mary Midgley's vision of our alienation from the material world. We now know enough about living organisms and the Earth System to see that we cannot explain them by reductionist science alone. The deepest error of modern biology is the entrenched belief that organisms interact only with other organisms and merely adapt to their material environment. This is as wrong as believing that the people of a village interact with their neighbors but merely adapt to the material conditions of their cottages. In real life, both organisms and people change their environment as well as adapting to it. What matters are the consequences: if the change is for the better then those who made it will prosper; if it is for the worse then the changers risk extinction. Reductionist science grew from the clockwork logic of Descartes. It can only partially explain anything alive. Living things also use the circular logic of systems, now more fashionably known as complexity theory, where cause and effect are indistinguishable and where there is the miracle of emergence.

President Havel's thoughts led me to think about the ethic that comes from Gaia theory; it would be one with two strong rules. The first rule states that stability and resilience in ecosystems and on Earth requires the presence of firm bounds or constraints. The second rule states that those who live well with their environment favor the selection of their progeny. Imagine sermons based on these rules. Consider first the guiding hand of constraint. I can see the nods of approval. People's own experience of the need for a firm hand in the evolution of their families and in society concurs with the evolutionary experience of Earth itself.

The second rule, the need to take care of the environment, brings to mind a sermon on the abominable transgression of terraforming—the technological conversion of another planet into a habitat for humans. What is so bad about terraforming is its objective to make a second home for us while we are destroying our own planet by the greedy misapplication of science and technology. It is madness to think of converting with bulldozers and agribusiness the desert planet Mars into some pale semblance of Earth when we should be improving our way of living with Earth.

The second rule also warns of the consequences of unbridled humanism. Early in the history of civilization, we realized that overreaching self-worship turns self-esteem into narcissism. It has taken almost until now to recognize that the exclusive love for our tribe or nation turns patriotism into xenophobic nationalism. We are just glimpsing the possibility that the worship of humankind can also become a bleak philosophy, which excludes all other living things, our partners in life upon Earth.

We have inherited a planet of exquisite beauty. It is the gift of four billion years of evolution. We need to regain our ancient feeling for Earth as an organism and revere it again. Gaia has been the guardian of life for all of its existence; we reject her care at our peril. We can use technology to buy us time while we reform but we remain accountable for the damage we do. The longer we take the larger the bill. If you put trust in Gaia, it can be a commitment as strong and as joyful as that of a good marriage—one where the partners put their trust in one another. The fact that they are mortal makes that trust even more precious.

Let us as scientists look more closely at the ethical and philosophical aspects of Gaia. I have put before you the proposition that, in addition to being a theory in science, Gaia offers a worldview for agnostics. This would require an interactive trust, not blind faith, and a trust that accepts that, like us, Gaia has a finite life span and is provisional.

37
CELEBRATING CREATION

CHET RAYMO

Even the sparrow finds a home, and the swallow a nest, where she rears her brood beside thy altars.

Psalm 84:3

Late last summer, in the west of Ireland, I spent a night in the Gallarus Oratory, a tiny seventh-century church of unmortared stone. It is the oldest intact building in Ireland, and one of the oldest in Europe. The oratory is about the size of a one-car garage, in the shape of an overturned boat. It has a narrow entrance at the front and a single tiny window at the rear, both open to the elements. Even during the day one needs a flashlight to explore the interior.

I can't say exactly why I was there, or why I intended to sit up all night, sleeplessly, in that dark space. I had been thinking about skepticism and prayer, and I wanted to experience something of whatever it was that inspired Irish monks to seek out these rough hermitages perched on the edge of Europe, or—as they imagined—the edge of eternity. They were pilgrims of the Absolute, seeking their God in a raw, ecstatic encounter with stone, wind, sea, and sky.

Chet Raymo's "Celebrating Creation" originally appeared in the *Skeptical Inquirer* 23, no. 4 (July/August 1999).

The Gallarus Oratory is something of a tourist mecca, but at night the place is isolated and dark, far from human habitation. From the door of the oratory, one looks down a sloping mile of fields to the twinkling lights of the village of Ballydavid on Smerwick Harbor.

The sun had long set when I arrived, although at that latitude in summer the twilight never quite fades from the northern horizon. It was a moonless night, ablaze with stars, Jupiter brightest of all. Meteors occasionally streaked the sky, and satellites cruised more stately orbits. Inside, I snuggled into a back corner of the oratory, tucked my knees under my chin, and waited. I could see nothing but the starlit outline of the door, not even my hand in front of my face. The silence was broken only by the low swish of my own breath.

As the hours passed, I began to feel a presence, a powerful sensation of something or someone sharing that empty darkness. I am not a mystical person, but I knew that I was not alone, and I could imagine those hermit monks of the seventh century sharing the same intense conviction of "someone in the room." At last, I was spooked to the point that I abandoned my interior corner and went outside.

A night of exceptional clarity! Stars spilling into the sea. And in the north, as if as a reward for my lonely vigil, the aurora borealis danced toward the zenith. How can I describe what I saw? Rays of silver light streaming up from the sea, as if from some enchanted Oz just over the horizon, shimmering columns of fairy radiance. As I watched from the doorway of oratory, I remembered something the nineteenth-century explorer Charles Francis Hall wrote about watching the aurora from the Arctic: "My first thought was, 'Among the gods there is none like unto Thee, O Lord; neither are there any works like unto thy works!' . . . We looked, we SAW, we TREMBLED."

Hall knew he was watching a natural physical phenomenon, not a miracle, but his reaction suggests the power of the aurora even on a mind trained in the methods of science. What then did the monks of Gallarus think of the aurora, 1,300 years ago, at a time when the supernatural was the explanation of choice for exceptional phenomena? Stepping out from the inky darkness of their stone chapel, they must surely have felt that the shimmering columns of light were somehow meant for them alone, a sign or a revelation, an answer to their prayers.

We have left the age of miracles behind, but not, I trust, our sense of wonder. Our quest for encounter with the Absolute goes arm in arm with our search for answers. We are pilgrim scientists, perched on the edge of eternity, curious and attentive. The Gallarus Oratory was built for prayer, at a time when the world was universally thought to be charged with the active spirit of a personal God: Every stone might be moved by incantation, every zephyr blew good or ill; springs flowed or dried up at the deity's whim; lights danced in a predawn sky as a blessing or portent. Today, we know the lights are caused by electrons crashing down from the Sun, igniting luminescence. But our response to the lights might still be one of prayerful attention, and they lead us, if we let them, into encounter with the Absolute.

Traditional religious faiths have three components: a shared cosmology (a story of the universe and our place in it), spirituality (personal response to the numinous), and liturgy (public expressions of celebration and gratitude, including rites of passage). The apparent antagonism of science and religion centers almost entirely on cosmology: What is the universe? Where did it come from? How does it work? What is the human self? What is our fate? Humans have always had answers to these questions. The answers have been embodied in stories—tribal myths, scriptures, church traditions. All of these stories derived from a raw experience of the creation, such as my experiences inside and outside of the Gallarus Oratory. All of them contain enduring wisdom. But as a reliable cosmological component of religious faith they have been superseded by what cultural historian and Roman Catholic priest Thomas Berry calls the New Story—the scientific story of the world.

The New Story is the product of thousands of years of human curiosity, observation, experimentation, and creativity. It is an evolving story, not yet finished. Perhaps it will never be finished. It is a story that begins with an explosion from a seed of infinite energy. The seed expands and cools. Particles form, then atoms of hydrogen and helium. Stars and galaxies coalesce from swirling gas. Stars burn and explode, forging heavy elements—carbon, nitrogen, oxygen—and hurl them into space. New stars are born, with planets made of heavy elements. On one planet near a typical star in a typical galaxy life appears in the form of microscopic self-replicating ensembles of atoms. Life evolves, over billions of years, resulting in ever more complex organisms. Continents move. Seas rise and fall. The atmosphere changes. Millions of species of life appear and become extinct. Others adapt, survive, and spill out progeny. At last, human consciousness appears. One species experiences the ineffable and wonders what it means, and makes up stories—of invisible spirits who harbor in darkness, of gods who light up the sky in answer to our prayers—eventually making up the New Story.

The New Story has important advantages over all the stories that have gone before:

- *It works.* It works so well that it has become the irreplaceable basis of technological civilization. We test the New Story in every way we can, in its particulars and in its totality. We build giant particle accelerating machines to see what happened in the first hot moments of the Big Bang. We put telescopes into space to look for the radiation of the primeval explosion. With spectroscopes and radiation detectors we analyze the composition of stars and galaxies and compare them to our theories for the origin of the world. Always and in every way we try to prove the story wrong. When the story fails, we change it.

- *It is a universal story.* Although originally a product of Western culture, it has become the story of all educated peoples throughout the world; scientists of all cultures, religions, and political persuasions exchange ideas freely and apply the same criteria of verification and falsification. Like most children, I was taught that my story—Adam and Eve, angels, miracles, incarnation, heaven, hell, and all the rest—was the "true story," and that all others were false. Sometimes our so-called true stories gave us permission to hurt those who lived by other stories. The New Story, by its universality, helps put the old animosities behind us.

- *It is a story that emphasizes the connectedness of all people and all things.* Some of the old stories, such as the one I was taught as a child, placed humankind outside of space and time, gifted us with unworldly spirit, and gave us dominion over the millions of Earth's other creatures. The New Story places us squarely in a cosmic unfolding of space and time, and teaches our biological affinity to all humanity. We are ephemeral beings, inextricably related to all of life, to the planet itself, and even to the lives of stars.

- *It is a story that asserts our responsibility for our own lives and the future of the planet.* In the New Story, no omniscient deity intervenes at will in the creation, answers prayers, or leads all things to a predetermined end. We are on our own, in the immensity of creation, with an awesome responsibility to use our talents wisely.

- *It is a story that reveals a universe of unanticipated complexity, beauty, and dimension.* The God revealed by the New Story is not the paltry personal projection of ourselves who attracted and bedeviled our ancestors. It is, in the words of the Jesuit theologian David Toolan, "the Unnamable One/Ancient of the Days of the mystics, of whom we can only speak negatively (not this, not that), a 'wholly other' hidden God of Glory," or in the felicitous phrase of novelist Nikos Kazantzakis, "the dread essence beyond logic."

We should treasure the ancient stories for the wisdom and values they contain. We should celebrate the creation in whatever poetic languages and rituals our traditional cultures have taught us. But only the New Story has the global authority to help us navigate the future. It is not the "true" story, but it is certainly the truest. Of all the stories that might provide the cosmological basis of contemporary religious feeling, it is the only one that has had its feet held to the fire of exacting experience.

The New Story informed my response to the dancing lights in the night sky at Gallarus. What I saw was not a portent or miracle, but rather nature's exquisite signature of the magnetic and material entanglement of Earth and Sun.

As the Sun brightened the eastern horizon and the last shreds of aurora faded, I was suddenly startled by a pair of swallows that began to dart in and out of the Gallarus Oratory, hunting insects on the wing. I followed them inside and discovered a nest with three chicks perched on a protruding stone just above the place I had been sitting. The mysterious presence I had felt so strongly in the darkness was not a god, nor spirit, nor succubus, nor demon, but the respirations and featherings of swallows.

38

SCIENCE AND RELIGION IN
AN IMPERSONAL UNIVERSE

MATT YOUNG

*If something is in me which can be called religious then it is the
unbounded admiration for the structure of the world so far as our
science can reveal it.*

—*Albert Einstein*

I used to have a colleague I shall call Robin. He is a bright guy and a good scientist, and I think highly of him. He is also a member of a small Baptist sect
and a biblical literalist. Once, Robin owed me a favor, so I said, in essence, "Sit
down. I would like to know why you hold your religious belief without evidence or, if you have evidence, what that evidence is."

We talked for the better part of an hour. Robin told anecdotes, talked
about reports of "miracles" from all over the world, and spoke of his inner conviction, his inner feelings. I asked why he thought the religion of his parents
was right and all others were (therefore) wrong. I asked if he would be a

Matt Young's "Science and Religion in an Impersonal Universe" originally appeared in
the *Skeptical Inquirer* 25, no. 5 (September/October 2001). It is adapted from his
book *No Sense of Obligation: Science and Religion in an Impersonal Universe* (1stBooks
Library, 2001).

Koranic literalist if he had been born in Islamabad instead of Cleveland. He called this my "accident of birth" argument, but had no real answer to it.

Early on, I asked whether his belief was allegorical, that is, an approximation to the truth, or simply his way of getting at God and no better or worse than someone else's. Was his belief a hypothesis that he would employ as long as it worked, or was it absolutely true?

No, he answered, it is absolutely true.

At the end of the hour, he said, as best I can recall, "Look, what you said earlier, about being a hypothesis. [Pause.] I guess it is sort of a hypothesis." Saying so made him feel threatened. You could see it in his body language, hear it in his voice, see it in his eyes. So I quickly stopped the conversation.

The discussion with Robin kicked off what has become a four- or five-year investigation into religion and the basis for religious beliefs. Specifically, I set out to demonstrate, first, that empiricism is the only way to establish reliable knowledge about the physical world and, further, to show why it is appropriate to examine the claims of religion empirically. Accordingly, I applied a scientific approach to claims made by religious believers and apologists. Whether or not the universe has a purposeful creator, after all, is a matter of fact. It is therefore inappropriate for people who generally support their beliefs with evidence to believe without evidence in God. What, then, is the evidence?

My investigation brought me from science and philosophy of science to religion and philosophy, biblical criticism, evolution and cosmology, mathematical physics, and the science of the brain. I do not have firsthand knowledge of many of these fields, so I have gone to the literature for my information. Except for a handful of books and articles on physics and one statistics paper, every one is accessible to the diligent layperson; that is, anyone could read the same material as I read and draw his or her own conclusion. I present mine here.

Contrary to postmodernist assertion, there is objective reality or, if you prefer, objective truth that exists independently of the observer and the belief system of the observer. I argue further that the only way to get at that truth—more precisely, the only way to approximate it, as a map approximates a continent—is through empirical observation. That observation must not be casual, however; observation must be supplemented with reason and care, or else you fall into related traps of believing what is agreeable to you and of relying on selectively chosen anecdotes or vague and unprovable hypotheses as supporting evidence.

The hypotheses of religion must be treated the same way as any other hypotheses: They must be examined critically and tested. That is, we must ask—we have an intellectual obligation to ask—are the hypotheses supported by the available evidence? In my book, *No Sense of Obligation* (1st Books, 2001), I have tried to show that they are not. I will give an all too brief summary of my conclusions here.

Hypotheses and Evidence

I have dismissed what I called "popular" beliefs such as the belief in signs or miracles on several grounds. First, most presumed miracles can be explained or accounted for without invoking divine intervention. Storms and other natural disasters are just those: natural disasters and not acts of God. We may therefore reject the arguments of those who give God credit for all that is good and ignore all that is bad; they are using evidence selectively in order to bolster a belief that they must intend to hold onto come hell or high water.

Similarly, we cannot accept the kind of wishful thinking that there must be a God because otherwise there would be no purpose to our existence, no fixed values, no universal code of morality. You cannot arbitrarily hypothesize, for example, a universal code of morality and then use the presumed existence of that code to "prove" that there must be a God. This hypothesis is not obviously true and requires evidence to support it. Basing one unsupported hypothesis on another, equally unsupported hypothesis is not progress.

Even though Bible codes are tantalizing because they appear statistically significant in a way that anecdotes do not, we cannot accept them as signs from God, particularly in light of the strong circumstantial case that the Bible was compiled from a multiplicity of sources (which are often at odds with one another). In addition, it now appears that the input data used to "uncover" the Bible codes may have been adjusted to achieve a desired result.

In the Western world, a great many people nevertheless think that the Bible is the literal word of God. The myriad errors and inconsistencies in the Hebrew Bible and in the Gospels ought to deliver a death blow to that belief: At most, the Bible is the word of God as interpreted and distorted by generations of oral tradition and then by later redaction. The Book of Jonah is so obviously a fiction that I am astonished any time I hear someone argue for its literal truth. The Gospels are not contemporaneous accounts of the life of Jesus, and they are unsupported by external evidence. Each successive account may be no more than an embellishment of the preceding account; only the first account is even roughly accurate, and there is no independent evidence for supernaturalism. As important as the Bible is, it is not the literal word of God.

Let us make a distinction between evil (that is, the deliberate infliction of harm on one human by another) and misfortune. Both are a problem for those who believe in a benevolent God. Evil, oddly, is less of a problem: You can argue that evil is an unfortunate but necessary side effect of our having been granted free will, but it is hard to justify debilitating diseases by the same argument. The Bible gives no answers to the problems of evil and misfortune. Specifically, the most commonly cited theodicy, the Book of Job, offers little or no help. The comforters mostly blame the victim, assuming that he has done something wrong, even though readers of the book know that he is a righteous man. God himself never once claims to be just: only powerful. He seems to be saying, "Might makes right," a sentiment that our society has long abandoned as moral justification.

We can, however, find a potential source of evil in biology. When we see analogies to evil in the animal kingdom, we are properly reluctant to classify them as evil. In our minds, only humans can perpetrate evil. I conclude, therefore, that evil does not exist except insofar as we define it. It needs little or no explaining unless we hypothesize a benevolent God. Indeed, the God hypothesis hinders our understanding of evil rather than helping it.

Philosophical Arguments and Evidence

I have therefore found wanting almost all the arguments of the laity and the clergy alike. How well do philosophers fare? Not well. Their proof texts are not as old as those of the scriptural literalists, but they seem as dated, and, except for a few philosophically minded scientists, philosophers of religion seem as unwilling to incorporate the discoveries of modern science into their worldviews as are the biblical literalists.

The Ontological Argument of Saint Anselm asks you to imagine a "greatest possible being" and goes on to argue that such a being must be real because existence in reality is "greater" than existence in imagination. The argument makes no sense to me. It is based on the unsound premise that any valid logical argument must necessarily apply to the physical world. Neither does it define greatness, so you cannot evaluate its comparison between greatness in reality and greatness in imagination. Finally, it is a wild extrapolation from the finite to the infinite, and it is not testable.

The Argument from First Cause argues that every event has a cause. It assumes that the universe cannot be infinitely old, so there must have been a first cause, which Thomas Aquinas identified with his preexisting notion of God. The Argument from First Cause fares slightly better than the Ontological Argument, but only because of the empirically supported claim that the universe has a finite age. If it has a finite age, then it probably had an ultimate cause. There is, however, no evidence that the ultimate cause was purposeful, so the Argument from First Cause ultimately fails as well.

The Argument from Contingency considers that all objects are contingent, that is, that objects exist only as a result of a series of past events that did not need to have happened. Some event or entity, however, created the universe, and that event or entity could not have been contingent, since its existence is based on no past events. The Argument from Contingency presumes that objects or events are contingent, rather than deterministic. It further presumes that the entity that created the universe was purposeful. Neither presumption is obviously true, and the Argument from Contingency fails for much the same reason that the Argument from First Cause fails: It assumes without evidence that the creation was initiated by a being.

The Argument from Design sees both design and purpose in nature and presumes therefore that the entire universe was designed for a purpose. As a general argument, it is weak, but a couple of modern variations are more compelling. One such variation, which I call the Argument from Evolution, is firmly grounded in the fact that complexity increases almost inexorably as (geological) time progresses. The haphazard nature of evolution, especially the periodic mass extinctions, however, argues strongly against the claim that the universe was created with intelligent beings or anything else as its ultimate goal. A related argument, the Anthropic Principle, argues that the universe is so "hospitable" to life that it must have been designed with life in mind. The Anthropic Principle seems to me to be completely circular and impossible to take seriously.

Another design argument, which I call the Argument from Mathematical Physics, depends on whether you think there is order at the deepest levels of reality. Even so, there is no a priori reason to ascribe such order to a purposeful creator, and the Argument from Mathematical Physics fails: The universe need not have been created by a mathematician just because we can describe it by mathematics.

The Argument from Religious Experience presumes that, if people tell you that they have had certain experiences, then those people should be believed. I was frankly surprised that professional philosophers take this argument seriously. There is not one shred of evidence, credible or otherwise, that mystical or religious experiences are objectively real and not hallucinations or other well-understood mental phenomena. That is, although the mystical experiences seem real, no one has ever devised a test that can be used to distinguish them from well-known and well-understood artifacts such as hallucinations and dreams. In addition, the religious experiences that people report are strikingly at variance with one another and highly dependent on the cultures of the reporters, which strongly suggests that they are mental phenomena.

Belief, Knowledge, and Feelings

I conclude that the evidence in favor of a purposeful creator, let alone a benevolent God, is so weak as to be virtually nonexistent. Indeed, it is so weak that we are justified in arguing that the God hypothesis has been falsified. There is almost certainly no purposeful creator and certainly no benevolent God.

What then do I believe in?

Believe is a strong word. I do not *think* that the universe had a purposeful creator. I am *almost certain* that God does not intervene in our affairs, that there is no absolute code of morality, and so on. I probably believe these things as firmly as all but the most rigid literalists believe the very opposite. I differ from the literalists, however, in my admission that I could be wrong and in my continuing search for the evidence, either way. In short, *I try to believe what I have to believe, not what I want to believe.*

I am nearly convinced that the universe is completely deterministic. Even if it is not, the wave function of a complicated quantum system such as a brain evolves with almost perfect predictability. Far more of our personalities may be determined by the physiology of our brains than is generally recognized. Indeed, my statement that the universe is deterministic compels me to hypothesize that all our actions and thoughts are determined once and for all by the laws of nature. In this sense we have no free will: Free will is an approximation that we make because we can do nothing else; it is a concept that we developed because we seem to be free and have a great many choices open to us. But I doubt that we are free in the strictest sense of the term.

Some people find this argument very threatening. It might imply that mind is an epiphenomenon, that is, the result of physiological processes in our brains and bodies, and nothing more. That there is no purpose to our existence. That one day there will be no more humans, no Earth, no universe as we know it. To me, however, these are plain physical facts with no moral or ethical content. The fact that we do not have immortal souls does not justify unethical behavior. We might like the world to be otherwise, but it is not.

The Cosmic Religious Feeling

What then can I propose in place of theism? First, the knowledge that the universe is intelligible. As a scientist, I see or read about phenomena that must seem like miracles to laypersons and certainly seemed like miracles to the ancients. The ancients postulated a god or gods to explain the natural order. Today, however, we find the universe understandable in terms of physical laws and have no need to invoke supernatural powers. In place of theism, I propose what Einstein called a *cosmic religious feeling*, an "unbounded admiration for the structure of the world so far as our science can reveal it." The awe and humility Einstein felt in the presence of the "magnificent structure" of nature were a genuinely religious feeling, but it was firmly grounded in reality and required no supernatural God.

Second, without a literal belief in a god who dictates moral codes or guides us along our paths through the universe, I propose the idea that we are grownups, on our own and responsible for ourselves, not children for whom someone else is responsible.

Finally, I offer, to those who want it, a religious humanism that is human-centered, not God-centered. In this view, our lives have meaning, but it is meaning that we and our communities give them, not meaning that is derived from a supernatural source. We have to act as if we had free will, because we can do nothing else. But we and our communities have to develop our own ethics. There are no moral imperatives and no universal code of morality, no automatic rewards for good deeds, no automatic punishment for bad deeds, no God looking over our shoulders. All we can do is strive to improve ourselves and our world, and we are completely on our own. Far from despairing, however, I consider hopeful the facts that medicine and sanitation have improved our health and longevity; science and technology have given us shorter working weeks, more abundant food and resources, and more leisure; and our political systems have given us more freedom and dignity. The power to improve the system further and to extend our good fortune to the rest of the world is in us and our own rational thinking, not in God. To put it in theological terms, we must seek our salvation in this world, because there is no other.

39

AFTERTHOUGHTS

PAUL KURTZ

I

Does contemporary scientific cosmology demonstrate the existence of God? Do the Big Bang, the anthropic principle, or the design argument justify the claims of theists? Skeptic inquirers deny that they do. The so-called singularity of the Big Bang does not tell us what happened before the initial bang, how or why it was caused. To infer a divine being as the cause of the universe only pushes our ignorance back one step; for one can always ask, Who or what caused God? This is an illegitimate question we are told by believers, but then is it not likewise illegitimate to seek a single cause or ground of being for the entire universe—to take a leap of faith outside of nature? If the existence of God is postulated to explain the physical universe, then what is the justification for the claim that He, She, or It is a *Person*, as theists assume; indeed, that such a person responds to our prayers?

Similarly the attempt by theists to invoke "intelligent design" has insuf-

Paul Kurtz's "Concluding Postscript" is adapted from his article "Are Science and Religion Compatible?" which originally appeared in the *Skeptical Inquirer* 26, no. 2 (March/April 2002).

ficient evidential support. Natural selection, genetic mutations, differential reproduction, and other natural causes are sufficient to explain the evolution of species without the interposition of design in the universe. The anthropic principle maintains that some form of "fine-tuning" is responsible for the existence of life, particularly human life, on this planet. But how does this accord with the extinction of millions of species as discovered in fossil remains? If one assumes a designer, what about conflict, malfunctioning, and evil in the universe? Why is this not evidence for unintelligent or bad design? It is the height of anthropocentric chutzpah to assert that the purpose of the fine-tuning of the universe is for the emergence of the human species!

Heidegger posed the question: "Why should there be something rather than nothing?" Skeptics doubt that this question is meaningful. How would one go about confirming the theistic answer that is proposed? Why not accept the *brute facticity* encountered in the world as the given: matter and energy, the variety and forms of things, events, qualities, and properties that we experience in nature, from electrons and atoms to planets and galaxies, from single-cell paramecia to dinosaurs, from daffodils to human beings, from social institutions to cultural expressions? Why not treat these in pluralistic rather than unicausal terms in the contexts in which we encounter them? A nonreductive naturalistic account of nature is open to the richness and diversity that is uncovered. Perhaps the appropriate response to these questions is that of the *agnostic*, who admits that he does not know the ultimate ground of reality (whatever that means). In any case, the skeptic finds insufficient evidence or reasons for the classical theist position, and in that sense he is a *nontheist* or *atheist.*

Those who defend the existence of the supernatural believe that a transcendent God is also immanent in the universe. If this is the case then his presence may be judged experientially. Skeptical inquirers have investigated the alleged paranatural evidence adduced for the existence of "discarnate souls," "near-death experiences," or "communication with the dead"; similarly for the "efficacy of prayer," the Shroud of Turin, and other alleged anomalous phenomena—which have been found evidentially lacking.

The presumption that religion offers a special kind of higher spiritual truth is thus unwarranted, as is the claim that there are *two* truths: those of science, justified on experimental and rational grounds; *and* those of religion, which transcend any empirical/rational confirmation.

The most reliable methods of inquiry attempt to satisfy objective standards of justification. The historic claims of revelation in the ancient sacred literature are insufficiently corroborated by impartial eyewitnesses and/or are based upon questionable oral traditions. These were compiled many decades, even centuries, after the alleged death of the prophets. Many miraculous claims found in the Bible and the Koran—for example, the claims of healings and exorcisms within the New Testament or the creationist account in the Old Testament—are totally unreliable. They depend upon the primitive science of ancient nomadic and agricultural peoples and cannot withstand scientific scrutiny.

Unfortunately, some proponents of these historic religions appeal to them in order to block scientific research. Freedom of inquiry is essential for human progress; any effort to limit science is counterproductive; for example, attempts to restrict embryonic stem-cell research on alleged moral-religious grounds. Opponents of such research argue that once a cell begins to divide (even if it grows to a small number of cells—a blastocyst), the soul of a person is implanted, and that any effort to experiment with this is "immoral." The postulation of the soul to prohibit scientific inquiry is reminiscent of the opposition to the findings of Galileo and Darwin. Insofar as religionists insist that they can issue an imprimatur or *fatwah* against the kinds of scientific research that they find offensive, the first conclusion that can be drawn is that there is a need for the strict *separation of religion and science*.

A second inference that may be drawn concerns the relationship between religion and morality. Stephen Jay Gould proposed that there are two *magisteria*, which he says do not compete and do not contradict each other. The domain of science deals with truth, he says, but that of religion deals appropriately with ethics. I think that this position is profoundly mistaken. Indeed I submit that there ought to be *a separation between ethics and religion*. I do not deny that religious believers have often espoused and supported moral behavior, including charitable and beneficent acts, love, sympathy, and peace. But many believers have also invoked commandments from On High, which they sought to impose on society. Moreover, religions have often disagreed about what are the basic moral imperatives; and they have waged warfare against other religious or secular moral systems. Religions often ground their moral commandments on faith and tradition and many have sought to oppose constructive social change. One needs to open ethical values and principles to examination in the light of rational and empirical considerations. Religionists have demonstrated that they have no special competence in framing or evaluating such moral judgments.

I say this because there is a vast literature in the field of secular ethics—from Aristotle to Spinoza, Kant, John Stuart Mill, John Dewey, and John Rawls. These thinkers seek to demonstrate that ethics can be autonomous and that it is possible to frame ethical judgments on the basis of rational inquiry. There is a logic of judgments of practice and valuation that we can develop quite independently of a religious framework. Moreover, science has a role to play in decision making, for it can expand the means at our disposal (technology), and it can modify value judgments in the light of the facts of the case and their consequences. The applied and policy sciences are normative insofar as they frame prescriptive judgments. Yet many people today mistakenly suppose that you cannot be moral without religious foundations. This is a false supposition; for ever since the Renaissance, the secularization of morality and the realization of naturalistic values continue quite independently of religious commandments.

A third area which has been hotly debated in the modern world is the relationship between religion and the state. Democrats vigorously defend the open secular society and *the separation of religion and the state.* Although religionists have every right to express their point of view in the public square, religion should be primarily a private matter. Religions should not seek to impose their fundamental moral-theological principles on the entire society. This would be tantamount to a theocracy. A democratic state is neutral, not seeking to favor one religion over any other or none. It does not seek to establish a religion; it does not legislate religious principles into law. In particular, it seeks to avoid the censorship of scientific inquiry.

II

What then is the proper domain of religion? Is anything left? My answer is in the *affirmative.* In a minimal sense, I think that religion and science are compatible, depending of course on what is meant by religion. Religions have performed important functions that cannot be easily dismissed. Religions will continue with us in the foreseeable future and will not simply wither away.

No doubt my thesis is controversial: religious language, I submit, is not primarily descriptive; nor is it prescriptive. The descriptive and explanatory functions of language are within the domain of science; the prescriptive and normative are the function of ethics. Both of these domains, science and ethics, have a kind of autonomy. Certainly within the political domain, religious believers per se do not have any special competence, similarly for the moral domain. It should be left to every citizen of a democracy to express his or her political views. Likewise, for the developing moral personality who is able to render moral judgments.

If this is the case, what is appropriate for the religious realm? The domain of the religious, I submit, is expressive and emotive. It presents moral poetry, aesthetic inspiration, performative ceremonial rituals, which act out and *dramatize* the human condition and human interests, and seek to slake the thirst for meaning and purpose. Religion—at least the classical religions of revelation—deal in parables, narratives, metaphors, stories, myths; and they frame the divine in anthropomorphic form. They express the existential yearnings of individuals endeavoring to cope with the world that they encounter and seek to wrestle some meaning in the face of death. Religious language in this sense is eschatological. Its primary function is to express *hope*. If science gives us truth, morality the good and the right, and politics justice, religion is in the realm of promise and expectation. Its main function is to overcome despair and hopelessness in response to human tragedy, adversity, and conflict—the brute, inexplicable, contingent, and fragile aspects of the human condition. Under this interpretation religions are not primarily true, nor are they primarily good or right or even just; they are, if you will, *evocative*, attempting to transcend contrition, fear, anxiety, and remorse, providing balm for the aching heart—at least for a significant number of people, but surely not all.

I would add to this the fact that religious systems of thought and belief are products of the creative human imagination. They traffic in fantasy and fiction, taking the promises of long-forgotten historical figures and endowing them with eternal cosmic significance.

The role of creative imagination, fantasy, and fiction can not be easily dismissed. These are among the most powerful expressions of human dreams and hopes, ideals and longings. Who can deny that so many humans are entranced by fictionalized novels, movies, and plays? The creative religious imagination analogously weaves tales of consolation and of expectation. They are dramatic expressions of human longing, enabling humans to overcome grief and depression.

In the above interpretation of religion as dramatic existentialist poetry, science and religion are not necessarily incompatible, for they address different human interests and needs. My interpretation of the function of religion is on the metalevel: classical religions are attempts to transcend the tragic dimensions of human experience, even though, in my judgment, they are false in what they claim or promise.

III

A special challenge to naturalism emerges at this point. *Methodological naturalism* is the basic principle of the sciences—we should seek natural causal explanations for phenomena, testing these by the rigorous methods of science. *Scientific naturalism,* on the other hand, goes beyond this, because it rejects as nonevidential the postulation of occult metaphors, the invoking of divine spirits, ghosts, or souls to explain the universe; and it tries to deal in materialistic, physical-chemical, or nonreductive naturalistic explanations. It is the second form of naturalism that especially worries theists, for it rejects their basic cosmic view. The frenzied opposition to Darwinism which persists today is clearly rooted in the gnawing fear that scientific naturalism undermines religious faith.

If this is the case, the great challenge to scientific naturalism is not simply in the area of truth but of *hope,* not of the good but of *promise,* not of the just but of *expectation*—in the light of the tragic character of the human condition. Darwinian evolutionists recognize that death is final, not simply the death of each individual but the possible extinction some day in the remote future of the human species itself. Evolutionists have discovered that millions of species have become extinct. Does not the same fate await the human species? Cosmological scientists indicate that at some point it seems likely that our sun will cool down, or indeed, looking into the future, that a deep freeze or big crunch may eventually overtake the entire universe. Some star trekkers, inspired by science fiction, romanticize that one day the human species will be able to leave the Earth, and inhabit other planets and perhaps other galaxies. Nonetheless at some point the deaths of the individual, the human species, even our planet and solar system seem likely.

What does this portend for the ultimate human condition? We live in an epoch where the scale of the universe has expanded enormously. It is billions of light-years in dimension. Much of this no doubt is based on mathematical and astronomical extrapolations, which may be altered by future science. Nonetheless, the role of the human species pales in significance in the vast cosmic scene. Clearly science enables us to explain much about our universe; and this knowledge is truly a marvel to behold. Yet the universe that we have uncovered far outstrips our small effects upon it. It was not made *for us,* as theists presuppose. Nor did God create us in his image; rather we etched him in ours. Does this naturalistic perspective of the universe and our place or lack of it within it forever extinguish any grandiose aspirations that humans harbor? Does it destroy and undermine hope? Does it provide sufficient consolation for the human spirit? The central issue for naturalists is the question of human *courage*. Can we live a full life in the face of our ultimate extinction?

These are large-scale questions, yet they are central for the religious consciousness. Can scientific naturalism, insofar as it undermines theism, provide an alternative dramatic, poetic rendering of the human condition, offering hope and promise? Countless numbers of brave individuals have lived significant lives and even thrived, aware of the possible future extinction of our species and our solar system. Many other humans apparently cannot bear that thought. They crave immortality; and theistic religion responds to their need. Others do not stay awake nights worrying what will happen five, ten, or fifteen billion years from now. They find life worthwhile for its own sake here and now.

There is another dimension of the human condition in the twenty-first century that we need to consider—the expanding immensity of the universe and the mystery of the unknown (Martin Gardner speculates about this in his essay in this volume). The dazzling reality of human imagination today is that our conceptions of the size of the universe are continually being revised and enlarged. This process has been going on at least since the sixteenth century. The Copernican revolution dislodged our terrestrial home as the center of the universe and replaced it with the heliocentric perspective. The twentieth century has seen several revolutions in cosmology: the theory of relativity, quantum mechanics, and the uncertainty principles that have altered our views of reality. But it is the Hubblean revolution that is of special relevance today because it has extended our perspective exponentially. It has taken us beyond our solar system and our own galaxy, the Milky Way, into the vast outer reaches of space—to other galaxies, nebulae, and black holes, where the birth and death of star systems and galaxies can be observed. The telescope named after the astronomer Edwin Hubble has left the density of the Earth's atmosphere and enables us to penetrate still deeper into the recesses of the universe. The idea of a constantly evolving and changing universe, not a static eternal one, now dominates the outlook of scientific cosmologists. In this we are like mere specks of dust on a minor planet at the edge of one galaxy hurtling through space. Perhaps there are other universes or multiverses, which from our perspective are virtually infinite in number. If one traverses in reverse from the macrorealm to the microrealm, one contemplates a similar virtually infinite number of subatomic particles. Thus our view of the universe is breathtakingly endless in either direction!

Many theories in contemporary cosmology are no doubt based on theoretical conjectures; these need to be verified before they are accepted. Nonetheless our present cosmic perspective heightens awe, astonishment, amazement. We are confronted by resplendent, elegant, and majestic scenes. Bertrand Russell once observed that insofar as the human mind is able to contemplate the vastness of the universe, it enlarges its horizons by becoming a citizen of the universe.

Recent discoveries of new planets in nearby star systems lead us to infer the probability that there are not only billions of stars and galaxies, but billions of planets. If we add to these discoveries the possibility, indeed likelihood, that other forms of life may exist in other parts of the universe, then we may not be alone. One might ask: Are not these reflections a form of religious contemplation? No, they need not be. They are, rather, forms of aesthetic exaltation at the beauty and scale of the cosmic scene, which Carl Sagan so admirably expressed in his television series *Cosmos*.

There are other attitudinal reactions that these speculations no doubt inspire. First, there is some appreciation of our own finitude and limitations. From this awesome perspective many human beings develop fear and trembling, dread and anxiety, perhaps even horror. No doubt a recognition of the limits of human power may lead many humans to supplicate the vast forces beyond and even to deify them. The familiar human tendency to develop anthropomorphic conceptions of the universe or to relate everything within it to human aspirations and hope is to give way to the transcendental temptation. In our dreams and reverie we are tempted to repeople the universe with Titans engaged in star wars. Our cosmological speculations here far outstrip the imaginative scenes of the science-fiction literature of the past.

There is, however, another possible attitude in response: the recognition of our power in our own sphere of action on our own planet and in our own solar system and our need to develop the courage to realize our plans and projects and to resolve to live fully. We can and do enjoy life, pursue science and reason, love and be loved, share our aspirations and hopes with other individuals, build a better global community, and prosper and exult as human beings.

Religious narratives in the last analysis are like other great works of art: Shakespeare's *Macbeth* or Verdi's *Requiem*, Michelangelo's *David* or Picasso's *Guernica*, Beethoven's Ninth or Mahler's Third symphonies. They dramatize various aspects of the human condition—though religious mythologies provide fanciful remedies that should not be taken as literally true. The question nonetheless remains: Can or should we seek to replace the mythological responses of monotheists with more realistic appraisals and proposals for coping with the human condition? Perhaps the beginning of wisdom is implicitly suggested: science has catapulted the human species into outer space, both in imagination and exploration; and it reveals a marvelous structure far more stunning than anything contrived by the ancient theistic religions. Can we as a species endure in a naturalistic universe, recognizing the elements of the mystery that still remains and the vast fringes of the unknown? Can we at the same time each find life worthwhile? Can we live with reason and courage, compassion and exuberance? That is the challenge facing scientific naturalists in the future, as science continues to enlarge our understanding of the cosmos.

CONTRIBUTORS

SIR HERMANN BONDI is a cosmologist and Fellow of the Royal Society, past Master of Churchill College, Cambridge University, and president of the Rationalist Press Association. He is a Laureate of the International Academy of Humanism.

VERN L. BULLOUGH, historian and sexologist, is distinguished professor emeritus at the University of California—Northridge and the University of Southern California. He is a fellow of CSICOP. Among his more than three dozen books are *Sexual Practices and the Medieval Church* (coauthor, Prometheus Books, 1984) and *Sexual Attitudes: Myths and Realities* (coauthor, Prometheus Books, 1995).

ARTHUR C. CLARKE is a well-known science-fiction author of such best-selling books as *2001: A Space Odyssey, Childhood's End,* and *Rendezvous with Rama.* He is a Humanist Laureate of the International Academy of Humanism.

RICHARD DAWKINS's latest book is *Unweaving the Rainbow* (Houghton Mifflin, 2000). He is an Oxford University zoologist, a CSICOP Fellow, a Humanist Laureate of the International Academy of Humanism, and a contributing editor to *Free Inquiry* magazine.

WILLIAM A. DEMBSKI, mathematician and philosopher, is associate research professor at Baylor University and a senior fellow with the Discovery Institute. He is author of *The Design Inference: Eliminating Chance Through Small Probabilities* (Cambridge University Press, 1998).

DANIEL C. DENNETT is director of the Center for Cognitive Studies at Tufts University. He is the author of *Darwin's Dangerous Idea: Evolution and the Meanings of Life* (Simon & Schuster, 1995), as well as many other works.

TANER EDIS is a professor of physics at Truman State University, Kirksville, Missouri. He is author of *The Ghost in the Universe: God in Light of Modern Science* (Prometheus Books, 2002).

JEROME W. ELBERT is professor emeritus of physics at the University of Utah. He is author of *Are Souls Real?* (Prometheus Books, 2000).

Nobel Laureate **RICHARD P. FEYNMAN** (1918–1988) was one of this century's most important physicists and critical thinkers. Feynman, who worked on the Manhattan Project, was a longtime professor at Cal Tech, where his lectures on physics were legendary. He is perhaps best known to the general public for his work on the committee investigating the *Challenger* explosion—another testament to his view that scientists have an obligation to apply their reasoning abilities to issues of importance to the general public.

ANTONY FLEW is professor emeritus of philosophy at Reading University. He is a Fellow of CSICOP and a Humanist Laureate of the International Academy of Humanism. Among his many books are *Merely Mortal? Can You Survive Your Own Death?* (Prometheus Books, 2001) and *Atheistic Humanism* (Prometheus Books, 1993).

KENDRICK FRAZIER is editor of the *Skeptical Inquirer*. His latest books are a new, updated edition of *People of Chaco: A Canyon and Its Culture* (W. W. Norton 1999), published in April, and *Encounters with the Paranormal: Science, Knowledge, and Belief* (Prometheus Books, 1998). He is the recent recipient of the "In Praise of Reason" Award.

MARTIN GARDNER, former veteran columnist with *Scientific American* and the *Skeptical Inquirer*, is author or editor of over sixty books. Among his latest are *The Colossal Book of Mathematics: Classic Puzzles, Paradoxes, and Problems* (W. W. Norton, 2001) and *A Gardner's Workout: Training the Mind and Entertaining the Spirit* (A. K. Peters, 2001).

OWEN GINGERICH is professor of astronomy, Harvard-Smithsonian Center for Astrophysics.

The late STEPHEN JAY GOULD, a recipient of the "In Praise of Reason" Award and a CSICOP Fellow, was the Alexander Agassiz professor of zoology and professor of geology at Harvard University. His most recent books are *Rocks of Ages: Science and Religion in the Fullness of Life* (Ballantine Books, 1999), and *The Structure of Evolutionary Theory* (Harvard University Press, 2002).

MORTON HUNT is the author of *The Story of Psychology* (Doubleday, 1994) and *The New Know-Nothings: The Political Foes of the Scientific Study of Human Nature* (Transaction, 1998).

PAUL KURTZ is professor emeritus of philosophy at the University of New York at Buffalo and the founding chairman of the Committee for the Scientific Investigation of Claims of the Paranormal, the Council for Secular Humanism, and the Center for Inquiry. He is author or editor of thirty-eight books. His most recent books are *Skepticism and Humanism: The New Paradigm* (Transaction, 2001), and *Embracing the Power of Humanism* (Rowman & Littlefield, 2000).

ANTHONY LAYNG is an emeritus professor of anthropology (Elmira College). He now lives in Winston-Salem, North Carolina. He has written numerous articles for *USA Today: The Magazine of the American Scene*, many of which have appeared in anthologies.

JAMES LOVELOCK is an independent scientist and inventor and is known as the progenitor of the Gaia Theory. He is a Fellow of the Royal Society and an honorary Fellow of Green College, Oxford. His most recent book is *Homage to Gaia: The Life of an Independent Scientist* (Oxford University Press, 2002).

TIMOTHY MOY is a professor of the history of science at the University of New Mexico.

JOE NICKELL is a senior research fellow of CSICOP. He is author of *Inquest on the Shroud of Turin* (Prometheus Books, 1983) and *Real-life X-files* (University Press of Kentucky, 2001), among other books.

DAVID NOELLE holds a faculty position at Vanderbilt University in the departments of computer science and psychology. He is active in cognitive neuroscience.

CIARÁN O'KEEFFE is professor of psychology, University of Hertfordshire.

BARRY PALEVITZ is a professor in the Department of Botany, University of Georgia, Athens. His interests include cell biology and science communication. He is coauthor of *Discovery: Science as a Window to the World* (Blackwell, forthcoming).

JACOB PANDIAN is professor of anthropology at California State University, Fullerton.

MASSIMO PIGLIUCCI is an associate professor of botany at the University of Tennessee. He is the author of *Denying Evolution: Creationism, Scientism,* *and the Nature of Science* (Sinauer Associates, 2002) and *Tales of the Rational: Skeptical Essays About Nature and Science.* He is a Fellow of CSICOP.

STEVEN PINKER is a professor of psychology and director of the Center for Cognitive Neuroscience at the Massachusetts Institute of Technology. He is a CSICOP Fellow and a Humanist Laureate of the International Academy of Humanism. His most recent book is *The Blank Slate: The Modern Denial of Human Nature* (Viking, 2002).

CHET RAYMO is author of *Skeptics and True Believers: The Exhilarating Connection Between Science and Religion* (Walker and Co., 1998) and other books such as *The Soul of the Night* (Hungry Mind Press, 1996) and *365 Starry Nights* (Prentice-Hall, 1982). He is science columnist for the *Boston Globe* and teaches physics and astronomy at Stonehill College in Massachusetts.

EUGENIE SCOTT, a CSICOP Fellow, is executive director of the National Center for Science Education, Inc., El Cerrito, California. She is a physical anthropologist, known for her defense of evolution and her criticism of creationism.

DAVID A. SHOTWELL has taught mathematics at the Colorado School of Mines, San Diego State University, and Sul Ross State University in Alpine, Texas, where he is now retired.

QUENTIN SMITH is professor of philosophy at the University of Western Michigan. He has published five books, including *Theism, Atheism, and Big Bang Cosmology* (Clarendon Press, 1993) with William Lane Craig. He is a philosophy editor at Prometheus Books and editor of *Philo*.

VICTOR J. STENGER is a professor of physics at the University of Hawaii. He is the author of *Not By Design: The Origin of the Universe* (Prometheus Books, 1988), *Physics and Psychics: The Search for a World Beyond the Senses* (Prometheus Books, 1990), and *The Unconscious Quantum: Metaphysics in Modern Physics and Cosmology* (Prometheus Books, 1995). He is a CSICOP Fellow.

IRWIN TESSMAN is a professor of biology. Purdue University, West Lafayette, Indiana.

JACK TESSMAN is a professor of physics emeritus, Tufts University.

NEIL DEGRASSE TYSON, an astrophysicist, is the Frederick P. Rose Director of New York City's Hayden Planetarium and a visiting research scientist at Princeton University. His recently published memoir is titled *The Sky Is Not the Limit: Adventures of an Urban Astrophysicist* (Doubleday, 2000). He is a CSICOP Fellow.

MATT YOUNG is adjunct professor at the Colorado School of Mines and retired physicist with the National Institute of Standards and Technology. He is the author of *Optics and Lasers* (Springer, 2000) and *The Technical Writer's Handbook* (University Science Books, 1989).

STEVEN WEINBERG is professor of physics and holds the Josey Regental Chair of Science at the University of Texas at Austin. His research has spanned a broad range of topics in quantum field theory, elementary particle physics, and cosmology, and he has been honored with numerous awards, including the Nobel Prize in Physics and the National Medal of Science. His books include the prize-winning *The First Three Minutes* (Basic Books, 1993), *Gravitation and Cosmology* (Wiley, 1972), *The Discovery of Subatomic Particles* (W. H. Freeman, 1990), and most recently *Dreams of a Final Theory* (Vintage Books, 1994). He is a CSICOP Fellow and a Humanist Laureate of the International Academy of Humanism.

RICHARD WISEMAN is professor of psychology, University of Hertfordshire and a research fellow of CSICOP and author of *Deception and Self-Deception* (Prometheus Books, 1997) and coauthor of *Guidelines for Testing Psychic Claims* (Prometheus Books, 1995).